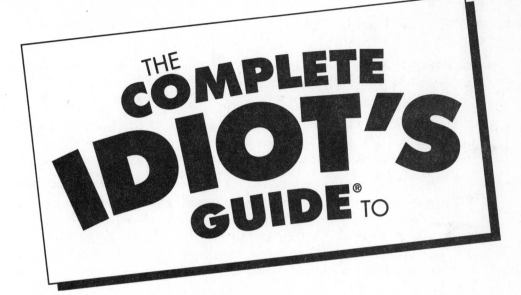

THE COMPLETE IDIOT'S GUIDE® TO

Building Your Own Home

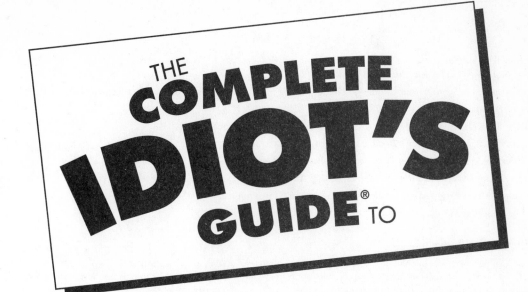

THE
COMPLETE IDIOT'S GUIDE® TO

Building Your Own Home

Second Edition

by Dan Ramsey

ALPHA

A member of Penguin Group (USA) Inc.

ALPHA BOOKS

Published by the Penguin Group

Penguin Group (USA) Inc., 375 Hudson Street, New York, New York 10014, U.S.A.

Penguin Group (Canada), 10 Alcorn Avenue, Toronto, Ontario, Canada M4V 3B2 (a division of Pearson Penguin Canada Inc.)

Penguin Books Ltd, 80 Strand, London WC2R 0RL, England

Penguin Ireland, 25 St Stephen's Green, Dublin 2, Ireland (a division of Penguin Books Ltd)

Penguin Group (Australia), 250 Camberwell Road, Camberwell, Victoria 3124, Australia (a division of Pearson Australia Group Pty Ltd)

Penguin Books India Pvt Ltd, 11 Community Centre, Panchsheel Park, New Delhi—110 017, India

Penguin Group (NZ), cnr Airborne and Rosedale Roads, Albany, Auckland 1310, New Zealand (a division of Pearson New Zealand Ltd)

Penguin Books (South Africa) (Pty) Ltd, 24 Sturdee Avenue, Rosebank, Johannesburg 2196, South Africa

Penguin Books Ltd, Registered Offices: 80 Strand, London WC2R 0RL, England

Publisher: *Marie Butler-Knight*
Product Manager: *Phil Kitchel*
Senior Managing Editor: *Jennifer Chisholm*
Senior Acquisitions Editor: *Mike Sanders*
Development Editor: *Lynn Northrup*
Senior Production Editor: *Billy Fields*
Copy Editor: *Tiffany Almond*
Illustrator: *Shannon Wheeler*
Cover/Book Designer: *Trina Wurst*
Indexer: *Aamir Burki*
Layout: *Ayanna Lacey*
Proofreading: *John Etchison*

For Florence Curtis Ramsey Ramos Delrow. Thanks for the typewriter!

Contents at a Glance

Contents

Foreword

In 1988, with a couple of major remodeling projects under our collective belt, we decided to build a house. What's more, we decided to contract it ourselves. Like you, we knew that the first step was to educate ourselves. We talked to anyone who would give us the time of day and would put up with our camera and microphones: architects, builders, developers, city inspectors and building officials, suppliers, subcontractors, lenders, real estate agents—all were fair game. It would have been a big help to us if Dan's book had been available then!

Whether you decide to be your own contractor or to hire the whole thing out, knowledge of all the processes involved—from planning and design to finance and construction—is vital to your success. You'd be wise to consult a variety of resources to make sure you get a full range of opinions and options. *The Complete Idiot's Guide to Building Your Own Home, Second Edition,* is an excellent place to start. It is comprehensive, accurate, and honest about all aspects of the building process. I predict your copy will be dog-eared and well worn by the time you finish building your new home!

At *Hometime* we have long held that self-contracting is not a job for the ill-informed or weak-of-heart. We think it is a job best suited to those who have the necessary skills to invest their sweat equity in order to spend their dollars on upgraded design and materials, or for those who simply have a passion for building— plus the time and money to fix the mistakes they will inevitably make. Self-contracting just to save money is not a good enough reason because, in our experience, most first-time builders will blow any savings they might have realized on fixing their errors.

However, we also think that anyone who is going to build a house, even as a passive but interested bystander, *must* have a solid working knowledge of all phases of the job. You must be able to effectively communicate your wants and needs to your design, construction, financial, and legal partners. Knowing the language and understanding the options is key to getting what you want at a price you can afford.

The other thing you need to know about building a house is that it can be a very emotional experience. After all, this is your *home* we're talking about, your safe haven from the world, the physical manifestation of your ideas about life and lifestyle. And that's the other reason you need to educate yourself. If you make home-building decisions based on emotion, your dream home will shortly become a nightmare. You have to be able to acknowledge emotion while relying on knowledge and intelligence to make commonsense decisions about the construction of your home.

We've gone on to build several more houses since that first one in 1988. And guess what? We're still making mistakes and learning new lessons. There's no immunity here—you will make mistakes, too. There are some things that only experience can teach. Fortunately, Dan Ramsey has a lot of experience and has talked to a lot of other people with a lot of experience. There's no rule that says you can only learn from your own experience—you can learn from others' experiences as well.

So before you take another step toward building your home, sit right down and read *The Complete Idiot's Guide to Building Your Own Home, Second Edition,* from cover to cover. Don't let the light, breezy style fool you—it makes the book a quick read, but there's a ton of good information in here. Dan starts with an overview of the entire process and then covers each step in more detail later in the book. You'll want to reread these chapters later as you approach each new phase of your project. And remember to consult other resources to round out your knowledge. Dan has made this easy for you by including a comprehensive list of reliable resources. If you make good use of this book and the resources it provides, you'll have a much better chance of having your project go smoothly. Chances are, you'll even enjoy the process!

Good luck with the house. I wish you many happy years in your dream home.

Dean Johnson

Hometime executive producer and co-host

Introduction

Building your own home used to be easy—anyone could do it. You simply gathered enough branches, animal skins, or ice blocks to protect you from the weather and wild animals. No permits or mortgage required. Easy!

Today, it seems like you need a permit to get a permit. Building your own home has become so complex that most folks settle for a prebuilt home that doesn't do more than protect them from the weather and wild neighbors.

You're not most folks. You want to direct the design and construction of a home that fits your needs and your wallet. You don't want a cookie-cutter house. You'd rather have a *real* recreation room rather than a converted garage. You want uniquely shaped rooms that make the home more open, rather than squared walls that make the carpenter's job easier. You want to take advantage of the view, rather than have a picture window that overlooks the neighbor's used car collection.

But you don't have all the time—and money—in the world. You want to plan and build your home on time and on budget. And you know that there are builders out there who will steal your time and your money if they have a chance.

So what's the solution?

Written by a home construction and remodeling contractor, *The Complete Idiot's Guide to Building Your Own Home, Second Edition,* is a concise, up-to-date guide on all aspects of deciding, designing, hiring, financing, building, and enjoying a home that's really your own. It answers every question you have about building your custom residence—and a few questions you may not have considered.

With clear instructions and illustrations, this book takes you through every step of the home-building process from figuring a budget to finding labor and materials to all aspects of home construction. Whether you plan to build it all yourself, build some and hire some done, or hire it all done to your specifications, *The Complete Idiot's Guide to Building Your Own Home, Second Edition,* will help you get the job done.

As important as getting the home you want, this book shows you how to save money along the way. How much money? Lots! In fact, you'll discover how to save up to 25 percent on the cost of building your home with the tips and ideas in this book. That's *tens of thousands* of dollars! You can use the savings to significantly reduce your mortgage or to add the amenities you thought you couldn't afford.

Thinking about someday building your own home? Want to do it the smart way? Want to know about housing alternatives? Start reading this book. It offers shelter from the storm.

Where to Find What You Need in This Book

Just like a quality home is built, this book follows a plan. Here it is:

Part 1, "Me, Build a Home?" introduces you to the *real world* of housing today. It answers important preliminary questions like these: How are houses built? How many are built by their owners? What are your housing options? How much can you save by building your own home? And many more topics.

Part 2, "Designing and Paying for Your Home," gets deeper into your quest for a home of your own. It covers designing your house, making it energy efficient, finding the best neighborhood and lot to build, estimating construction costs, and—very important—how to finance your new home.

Part 3, "Hiring Some Help," gives you the help you need to get it built. You'll learn about building codes and permits, hiring a contractor, hiring subcontractors, being your own contractor, and getting the best prices on materials and services.

Part 4, "Constructing Your Home," is big! This is your step-by-step guide for building a home of your own. You'll learn how to plan for success, and then tackle every task from preparing the building site to decorating the interior.

You'll also find two appendixes: a comprehensive glossary of nearly 400 construction terms, and a listing of helpful resources, including web addresses. Everything has been reviewed and updated for this new revised edition!

Extras

As if all that wasn't enough, you'll find lots of other tips, techniques, asides, and miscellaneous tidbits set off in the following sidebars:

Building Your Vocab

Ever feel like there's a secret society of homebuilders and they've all conspired to develop their own dictionary? Well, forget that! We're going to break building code with clear definitions of words you'll hear as you build your own home. Except for the expletives.

On the Level

No kidding! Really? In these boxes you'll find some cool bits of information that you can quickly read to keep up your interest.

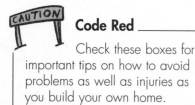

Code Red

Check these boxes for important tips on how to avoid problems as well as injuries as you build your own home.

Ka-ching!

Hear that cash register ring? You won't as much if you follow the proven money- and time-savers offered in this box.

Acknowledgments

I could probably fill an entire chapter with the names of people who have helped contribute to my construction career and to the development of this book. But I won't. Instead, here's a long paragraph ….

Thanks to Charlie House, who first showed me how to build—and grinned as I lost *two* thumbnails! And to Stan Eoff and other contractors who guided my learning experience. I'm also very grateful for the help and friendship of Greg Dunbar, a superlative contractor (and a great guitarist!). Writing and revising this book, I received help and resources from:

Ken Katuin of Abbisoft House Plans (www.homeplanfinder.com)

Dan Reif of Design Works (www.homeplanner.com)

James Mcnair and Chris Berrien of Garlinghouse (www.garlinghouse.com)

Don Mann of Grandy Post & Beam Homes (www.postbeam.com)

Bill Farmer of Mendo Mill, Willits, CA; and Bob Doty of Mendo Mill, Ukiah, CA

Robert Tyrrell, Heath Dowdy, Bob Wilson, and Jerry Goodrick of Hertz Big 4 Rents, Ukiah, CA

Scott Harris of ART, Inc., developers of Chief Architect for loaning me their professional design software. It's a great tool!

And thanks to Made E-Z Products, LLC (www.MadeE-Z.com), developers of "Contractors' Forms Made E-Z."

Thanks also to my hard-working agent, Sheree Bykofsky; to Mike Sanders of Alpha Books, whose vision made this book possible; to Marie Butler-Knight, who offered numerous improvements to this revised edition; to Lynn Northrup, Billy Fields, and Tiffany Almond, whose professional efforts enhanced this book; and to my partner and best friend, Judy Ramsey, who inspires me to build.

Special Thanks to the Technical Reviewer

The Complete Idiot's Guide to Building Your Own Home was reviewed by an expert building contractor who double-checked the accuracy of what you'll learn here, to help us ensure that this book gives you everything you need to know about building your own home. Special thanks are extended to Greg Dunbar. Greg has more than 20 years of construction and remodeling experience as a general contractor and subcontractor.

Trademarks

All terms mentioned in this book that are known to be or are suspected of being trademarks or service marks have been appropriately capitalized. Alpha Books and Penguin Group (USA) Inc. cannot attest to the accuracy of this information. Use of a term in this book should not be regarded as affecting the validity of any trademark or service mark.

Part 1

Me, Build a Home?

You've been peeking at home design books for years, lustfully wishing you had the courage and the cash to build your own dream home. So what's stopping you? For most folks, it's the fear of building a monstrosity the neighbors will dub "Frankenstein's Castle." Or maybe it's the fear of running out of money before you run out of plans. Or spending so much time pounding nails that you lose your day job.

Not to fear, Ramsey's here! Based on my own experience as a home construction and remodeling contractor—and with the help of scores of smart friends—I'll show you how you can participate in building your own home. Maybe, for you, that means building the whole thing yourself. Or perhaps you'll opt to be the boss and hire experienced tradespeople to do the real building. Or possibly you'll hire yourself as a sidewalk supervisor who isn't going to let the contractor get a dime he or she doesn't earn.

In any case, this part will get you started toward your goal. It includes a realistic look at housing today and how more than 100,000 houses are built each year by their owners. It also includes a peek at alternatives including kit, log, timber, post-and-beam, and manufactured homes. Maybe one of these is a better choice than a "stick-built" house. Finally, it will show you how to get more value for every dollar you spend on your dream home.

So stop dreaming! You can do this!

Nailing the *Real* World of Housing

In This Chapter

- ◆ Who builds their own home and why
- ◆ Starting to plan your new home
- ◆ How much of a home can you afford?
- ◆ Surviving the construction of your home

The claims made about building your own home sometimes sound like an infomercial: "Yes, you, too, can save as much as 35 percent *or more* on the construction of your new home. Imagine … an opulent residence second only to Buckingham Palace for just a few dollars a day—and the pride of gloating, 'I did it myself!'"

Hey, we're not that naive! The truth is that the guy's house cost 50 percent more than he budgeted and can never be resold because it has "design issues." And he lost his day job because he was sneaking away to pound nails.

So what's the *truth* about building your own home? Here are the facts. Can you build it yourself? Probably. Can you save money? Probably. Will you get a unique home? Probably. Will you be proud of what you've accomplished? Probably.

Pretty definitive, eh?

Actually, this chapter will help you turn that "probably" into a "yes" or a "no." How? It offers an honest look at the real world of housing today and how owners can participate in the building process.

The Truth Will Set You Free!

Okay. Let's get some facts out on the table. First, according to folks who know (read: the Census Bureau and the Department of Housing and Urban Development), the following are true:

- ◆ 1,386,300 new single-family homes were built in the calendar year 2003.
- ◆ 123,000 (8.9 percent) of these homes were built by the owner.
- ◆ Another 195,000 homes (14 percent) were built by contractors on the owners' land.
- ◆ The rest of them were built for sale.

Those nice folks who keep all the statistics also tell us that the typical (median) SFR (single-family residence) built today is 2,320 square feet in size. That's nearly 11 percent larger than just a decade ago.

How much does this "typical" house cost? That depends on where you live. In expensive areas (you know the ones), "typical" means $400,000 to $600,000. But spread out across the nation, typical new SFR value is more like $225,000.

On the Level

In the United States, home values have increased 36 percent on average over the past five years, nearly three times the rate of inflation for the same period, report economists. Topping the list is San Jose, California, where prices have increased more than 120 percent in the last half-decade.

What kinds of homes are people building? You name it: conventional lumber homes; less conventional timber, pole, beam, and even adobe homes; log homes; homes from kits; modular or manufactured homes; and others. (Chapter 2 offers a look at how conventional homes are built. Chapter 3 considers all the other types.)

Who builds their own homes? Most anybody who wants to. Some people tackle everything from cutting the trees and milling the lumber to hand-splitting shakes for the roof. Most participate to lesser degrees, ranging from sidewalk supervisors to subcontractors. There are about 30 separate task groups (lot preparation, framing, plumbing, and so on) in the construction of a typical house. (Chapter 12 tells you how to hire a contractor; Chapters 13 and 14 show you how to hire or be your own subcontractor. You're covered!)

The people who build their own homes range from couples living off the land (or a rich relative) to fully employed folks who manage the construction to retired people who take on some but not all of the work.

Can *you?*

Probably.

About 25 percent of all new homes are built by or for the owner.

(© homeplanner.com)

Why Build Your Own Home?

Hey, building a house is *work!* Why should you even consider building one? There are lots of good reasons—and good excuses. Here are the top contenders:

♦ Save money
♦ Pride of ownership
♦ Custom design
♦ Unconventional construction

Let's take a closer look at each of these reasons.

Saving Money

"But I thought the idea of saving money by building my own home is inaccurate," you say. Not really. You actually *can* save money as you build your own home. However, most folks don't. Depending on how much you participate in the project management and construction, you can theoretically save the general contractor's fees, an architect's fees, and maybe some or all of the costs of labor.

But most *owner-builders* either lose other income (such as from a job) or put the savings back into the house as upgrades. A smart owner-builder actually gets more for her or his money—but spends about as much. That's the reality of building your own home.

Building Your Vocab

The **owner-builder** is the general contractor, owns the land, may hire subcontractors for some or all of the building, and intends to occupy the house.

Bragging Rights

U.S. naturalist and author of *Walden* Henry David Thoreau built a cabin in the woods of Massachusetts in 1845, at Walden Pond. And he earned the right to brag about it. "I never in all my walks came across a man engaged in so simple and natural an occupation as building his house."

Henry's out-of-pocket cost was less than $30 for nails and other unnatural things. Most of the materials came from the land itself. Pride of ownership seems to go deeper with those who do some or all of the building of their home.

That's okay. Building your own home gives you the right to take pride in its design and construction. Even if your participation is in offering ideas to the designer and keeping the contractor on track, you are participating more than you would be with an I'll-take-that-one house.

On the Level

Architects' fees typically range from 8 to 15 percent of the home's value. In addition, about 25 to 50 percent of the price of a single-family residence goes for the land and utility hookups.

This book will show you which path to choose toward building the home you want.

Personalizing Your Home

A very popular reason why folks build their own home is to make it more like what they want. Cookie-cutter homes are pre-built for the "typical" homebuyer. Even plan-service plans typically don't have the personal amenities that can make a house feel like *your* home.

Code Red _____

Contractors typically require 4 to 6 months to build a home. Owner-builder homes typically take 9 to 24 months to build, so plan accordingly.

Building your own home helps you make it more personal. It may be as simple as modifying the function of a couple of rooms or adding a few windows to a stock plan. Or it may be adding various upgrades that enhance the usefulness and value of the house. Or it can be as far-reaching as designing the entire house from your ideas. You might not do the actual construction, but your personalization makes it "your home."

Are you asking for trouble if you want to design your own house? Not at all! In fact, most owner-builders and those who have others build their home do so from their own designs. You can do this. Chapter 5 will show you how—including how not to design your house.

Building a Unique Home

Some folks would rather not live in a "conventional" home. They prefer less conventional building materials (logs, timbers, bales, stone) or styles (A-frame, dome, yurt). Unfortunately, it may be more difficult to find contractors willing to build your dream home. So, you may have to do it yourself—or at least be the project manager.

Still, uniqueness is a great reason to build your own home. Many thousands of homes have been built for this reason alone.

You *Can* Build Your Own Home

Let me qualify that statement by saying this: You can participate in some or all of the design and construction of a home, subject to some commonsense limitations.

What kind of limitations? Stuff like having enough money to finish the project, having time to do it, having a place to stay until the house is done, being in good health, and knowing or picking up the skills you'll need. Commonsense stuff.

Not to worry. This book covers the money, the time, the meantime, and the requirements for building your own home. You'll learn how much money you'll actually need, how to decide which jobs to tackle yourself (and which to let someone else tackle), and how to get the skills needed. Best of all, you'll be getting *real-world* advice rather than I'm-selling-something platitudes. You'll also gain from my experience as a licensed general building contractor.

So let me answer one of the first questions that many folks ask as they consider building their own home: What's the trick?

The answer is that you can save some money and (very importantly) get more of the home you want by being your own general contractor (GC). A GC or building contractor is the boss. He or she is a licensed expert who knows how houses are built and *hires* subcontractors (subs) to build them. That's right! GCs don't build houses! They *manage* the construction of houses. For their knowledge and time, they get between 10 percent and 20 percent of the total cost of the project (including land). That means they get $25,000 to $50,000 for building a $250,000 house.

But wait, there's more. A general contractor may also get "fees" from subcontractors ("kickbacks" is such a nasty word) who are selected. GCs will probably get all materials for the job at a discount, some or all of which they keep. Some unscrupulous contractors also make money by not installing things as purchased. For example, one grade of subflooring is selected but a lower grade is installed. (I'll show you how to beat that con.)

So a general contractor can get 30 percent or more of the total cost of your home. Or *you* can get some or all of it. That's the trick to building your own home.

Lots of books out there will tip you off to this "trick." This book actually shows you how to hire GCs who don't pad bills or require fees from subcontractors. It shows you how to be your own GC and save their fee. Most important, it shows you how to put your savings into upgrading your home into a dream home.

Part 3 tells you how to hire and manage good folks who will do the job right, on time, and still make a reasonable profit for themselves. Most important, it offers *real-world* advice on how to make home construction decisions.

Ka-ching!

The best time to build your own home is when contractors aren't as busy. Why? Because busy contractors want more money and will take more time to build your home. In addition, subcontractors are more difficult to find and hire when they are tied up with busy general contractors. Makes sense. How can you find out if contractors are busy? Ask the contractor sales clerk at a large building materials supplier. They can help you find available and qualified contractors.

John and Marilyn Stout built their own home almost entirely themselves. John, a manufacturing supervisor, used his skills as a manager to be the house's general contractor. Marilyn took classes at a nearby community college on home construction and did much of the site work. John joined her on the weekends.

Richard and Florence Delrow decided on a manufactured home for retirement. Their daughter-in-law worked with the manufacturer on the design and decorating. Their son handled negotiations and oversaw construction and placement, keeping the job flowing.

Appendix B includes numerous resources for building your home including books, classes, videos, consultants, suppliers, and so on.

What Do You Want?

So now that you know that you can probably build your own home, the question arises: What do you want in a home? Fortunately, Part 2 covers that topic thoroughly. You'll get ideas on designing your home, making it energy efficient, selecting a location and building site, estimating construction costs, and paying for it all.

For right now, start making notes on what you want. In fact, start your Home Book. This can be a loose-leaf binder, a steno pad, or files on your computer. Whatever. It's a place to jot down ideas, contacts, resources, and other important information as you figure out what you want in your own home. Think about houses you've lived in or visited. What did you like or not like about each? How long do you hope to live in the house, and how will your needs change during that period? Jot down these thoughts in your Home Book.

Here are some other considerations:

- ◆ Size and complexity of the home
- ◆ Cost of the home vs. your financial assets
- ◆ Your mechanical skills and confidence
- ◆ Your people skills

On the Level

Three out of four self-built homes are constructed outside of metropolitan areas. Why? Because building your own home is typically easier in smaller towns and rural areas where building codes aren't as stringent. In addition, many people who build their own home want acreage.

◆ Your experience

◆ Your available time

Again, all these topics will be covered in this book. For now, keep your mind open to new ideas. You might not have any building experience, for example, but you might be able to take a working vacation at an owner-builder school. Or you might find ways of increasing your financial assets. Start the thinking process—and remember to make notes in your Home Book.

Do You Have the Money?

"Is there anyone here who, planning to build a new house, doesn't first sit down and figure the cost so you'll know if you can complete it? If you only get the foundation laid and then run out of money, you're going to look pretty foolish. Everyone passing by will poke fun at you. 'He started something he couldn't finish.'" (Luke 14:28–30 in *The Message: New Testament in Contemporary English*, Eugene H. Peterson; NavPress, 1993.)

On the Level

General contractor Greg Dunbar advises that one of the greatest skills a contractor or owner-builder needs is visualization. You must see a plan or a change in your mind to avoid costly mistakes. Another tip he offers is to plan hard and work easy. It's difficult and expensive to "erase" mistakes in home construction. Putting lots of time in the planning stage makes the actual job easier and less expensive.

The financial side of building is covered in Chapters 9 and 10. In the meantime, how can you make a quick estimate of the money it will take to build your own home? Just a ballpark figure for now? Start shopping for a similar home and location. Watch the newspapers, talk with local real estate agents, go to open houses. In a short time, you'll find out what homes cost both in price and in payments.

Factor in that most owner-builders upgrade their homes to be better than average—also known as more costly. If your savings match your additional costs, you come out even. But you may want to add 10 percent or more to comparable home values to compensate for your rich tastes.

Also consider hiring a local real estate appraiser to give you some figures on what your new home will cost. They won't want to give you detailed numbers until you know more about what you'll be building, but they can supply some local rules of thumb, such as a per-square-foot value, based on experience.

You might also hire a real estate agent to help you build your home. Be upfront with the agent about what you're trying to do, and she or he will probably get comparative values, lot costs, and financial guidelines for you. Why? The agent may wind up selling you a lot—or a home, if you decide not to build on your own.

On the Level

Home values vary greatly across the United States. A four-bedroom, two-bath, 2,200-square-foot home in Eau Claire, Wisconsin, is $135,000. The approximately same home in Palo Alto, California, is $1,400,000—more than 10 times greater. Even within a state, prices can vary. In Denver, Colorado, the price averages $250,000, while a similar home in Colorado Springs is valued at $160,000, according to industry sources.

You now know the general price range of homes you realistically like. How can you figure monthly payments?

First, figure out how much you're going to have to borrow. That is, deduct the value of any of your assets (home equity, savings) from the price of the home. Let's say the home you want is about $250,000 and you have about $15,000 to put toward the house. That's probably enough for a 95 percent loan. You'll need to borrow about $235,000.

Next, ask a lender or real estate agent for the current cost-per-thousand rate on a 30-year mortgage. That's the monthly cost for every $1,000 borrowed. It will be somewhere between $6.50 and $7.50 per thousand. In the example, multiply 235 by, say, $7.00 and you'll get $1,645 for principal and interest (PI) on a mortgage. Most mortgage companies are going to want you to make about three times that amount ($4,935, in our example) a month to qualify for that loan. (We're assuming good credit here.)

So, going the other way, take your monthly income and divide it by three, and then divide by the cost-per-thousand, and you have the amount you can mortgage. Add in your assets, and you know about what you can afford for housing. Keep in mind that these calculations are *very* general and can only give you a rough idea of what home you can afford. Chapter 10 offers a much more accurate way of calculating how much home you can afford.

What this rough calculation *does* tell you is that saving money on construction can reduce your mortgage or, if you're like most folks, increase the value of your home. You can get a custom home for the price of a standard home if you're willing to participate in construction.

On the Level _____

About 20 percent of all self-built homes are financed by cash, compared with only 7 percent of all single-family homes.

Surviving the Process

There are more costs to building a house than just financial ones. Pounding nails when you should be at your regular job can cost you your regular job. Working all day at your day job, and then framing or plumbing all night can cost you your health. Ignoring your spouse, family, and/or friends can cost you your relationships. Most important, doing something that is way over your head can cost you your head. You don't want to build your dream home only to be put in a state home.

What's the solution? Knowledge! Knowing yourself, your employer, your health, your relationships. And knowledge of the home designing and building processes. You're in charge of the "know thyself" stuff. This book will bring you up to speed on designing and building a home you will be proud to live in.

Code Red _____

What's the biggest and most costly mistake that owner-builders make? Overbuilding. They make costly changes during construction, or they build a home that they cannot profitably sell. Have a plan, study it carefully, and stick to it!

Now that you know the basics of what's involved in building your own home, it's time to grab your tool belt! Chapter 2 shows you how conventional homes are built.

The Least You Need to Know

- Every year, more than 120,000 conventional, kit, log, and manufactured homes are built by owners just like you.
- Chances are you can build your own home, or at least participate in building it, to get more for less.
- Building your own home means you'll get more of what you want for the same money.
- You can save thousands of dollars by participating in the construction of your house—money you can use to upgrade your home.
- With smart planning and good advice, you *will* survive the building process!

Sticks and Stones: How Homes Are Built

In This Chapter

- Understanding your home's plan set
- Working with general contractors and lenders
- Watching as your home is built from the ground up
- How to deal with building permits and inspectors
- The final touches that make a house a home

Before you decide how to participate in building your own home, let's spend some time at a typical residential building site. It's great fun!

In this chapter, you'll see how a contractor and crews prepare the building site, build the foundation or basement, add the floor, put up the walls and cover or sheath them, add roofing and siding, and then go inside and finish things up. Later chapters will expand on each of these tasks, offering comprehensive instructions so that you can do it, manage it, or double-check it, depending on how much of it you decide to do.

So get your grungy clothes on, lace up your boots, pack your lunch pail, grab your boom box, and let's head out to the construction site.

Rolling Out the Plans

It's 7 A.M. and the general contractor who was supposed to meet you here by now hasn't shown up. You hired the GC to advise you on construction of your house. Where the heck is he?

You walk over to a sawhorse table set up and see that he must have been here previously and has unrolled the large sheets of plans. Lots of sheets. They're labeled, so let's take a look at them:

- Site plan
- Foundation plan
- Exterior elevations

◆ Floor plan

◆ Framing plan

◆ Wall section

◆ Interior elevations

◆ Door and window schedule

◆ Room finish schedule

◆ Specifications

Wow. Lots of paperwork!

Typical elevation plan.

(© garlinghouse.com and homeplanfinder.com)

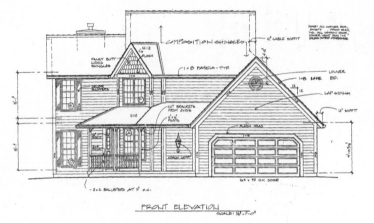

Typical floor plan.

(© garlinghouse.com and homeplanfinder.com)

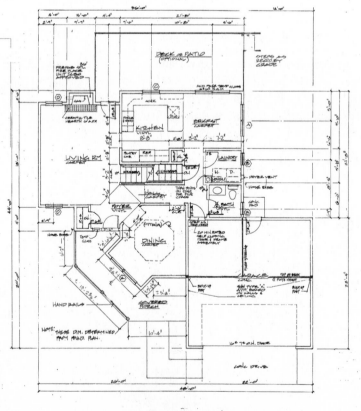

Just then, your GC drives up, steps out of his pickup, and crosses the vacant lot to greet you. "Sorry I'm a few minutes late. I got looking at the plans earlier this morning and noticed that I had forgotten the mechanical plan. That's the one that shows the heating and air-conditioning subcontractor what to do. Pretty important."

Who's the Boss? General Contractor

You've hired this GC to advise you through the maze of paperwork and decisions required for building a house. A general contractor or building contractor is actually a construction manager. As you learned in Chapter 1, the GC may never pound a nail or wire a switch. He or she manages those who do. If you are your own GC, it will be your job to make sure things get done when they should.

The plans you discovered are the directions for subcontractors and tradespeople to follow in building your house. Certainly, they know their trade—how to hang a door or install cabinets, for example—but they don't know how *you* want it to look when it's done. So they refer to the appropriate plans.

The general contractor hires subs and tradespeople who can read the plans accurately and understand what's needed before it is built. The GC should also check the finished job to make sure it conforms to the plans.

 **Code Red**

The plans for your home should be exactly what you want, not subject to interpretation. Otherwise, you could have an electrical box installed behind the bathtub!

Now you've read this book all the way through, purchased a full set of home plans from an architect or plan service, and decided that you can be your own GC. But just to make sure, you've hired an experienced builder to answer questions. Once a week the builder meets with you at the site to discuss any problems and answer any questions. His service isn't free, but it's cheaper than hiring a full-time GC for your house.

If you're confident about this house-building thing, you may decide to hire one of the subcontractors to advise you on construction. Typically, it's an experienced framing contractor. If you're really adventurous, you can use one of the online consulting services to advise you via the Internet and e-mail. If you're building in an area where few contractors are available, this can be the best option.

Anyway, Chapter 12 will give you a much closer look at what the general contractor does for his or her money. And Chapter 13 outlines how you can be your own GC.

The Unseen Lender

How are you paying for all of this? Unless they have a rich uncle, most folks get a mortgage. Chapter 10 will guide you through the process of finding a lender, getting a construction loan, and then converting it into a mortgage. In the meantime, let's discuss how the lender earns their money from the comfort of an office.

The lender can be a banker, a savings-and-loan officer, a mortgage broker, or some other financial professional who has experience with lending on new construction projects. It's helpful if the lender also has experience with owner-builders.

First, remember that the lender doesn't want your house. They especially don't want your partially built house. What they do want is to make sure of the following:

1. The house will be finished and livable (and resalable).
2. The workers will get paid (and not sue anyone).
3. You will pay for the house.

So the lender will want to get credit information on you, make sure the plans are buildable, and verify that the contractors are reputable. To keep everyone honest, the lender will hang on to the money until the required work is done.

Construction loans typically have a draw schedule. That means the electrical subcontractor receives or draws payment only after his job has been verified as successfully completed. How does the lender know what's done? An experienced lender may visit the site when called and verify completion at each stage. Most lenders hire staff or self-employed inspectors to do the job. Or they wait for the local building inspector to approve it.

There's lots more to know about lenders and construction loans. It's coming in Chapter 10. For now, let's assume that some wise lender has approved your construction loan subject to a draw schedule. You can now hire the subs needed to build your own home.

Ready: Site Prep and Foundation

You've already been down to the local building department and filed for all the permits you'll need (see Chapter 11). Good for you. Now it's time to start getting your land ready for your house. It's called site preparation, or prep, and includes removing any obstructions (trees, boulders, old buildings), leveling out the land, and getting everything ready for the house's *foundation*. It also means planning for drainage and water runoff from the site.

Building Your Vocab

The underlying support for a structure, including footings, walls, and piers, is called the building's **foundation**.

The vacant building lot in our example was recently purchased, so it still has flagged corner stakes—that is, wooden stakes with colorful ribbons that are stuck in the ground to mark the corners of the property. Fortunately, you bought a level lot, so little prep is needed except to check the plans and put some more stakes in the ground.

The first stakes mark utilities such as where the sewer line comes in from the street, the sewer tap. If your lot were in the country and you were installing a septic system, your site plan would show you where to mark out the tank and field. The next set of stakes mark the final location of the actual building(s). This is where the foundation of your home, garage, and any other foundation structures will go. You'll need to do lots of measuring, because you don't want to move the house if it's in the wrong place. The plans call for your home to have "setbacks," or designated distances between the structure and property lines.

Chapter 17 will guide you step-by-step through site preparation.

Typical foundation layout.

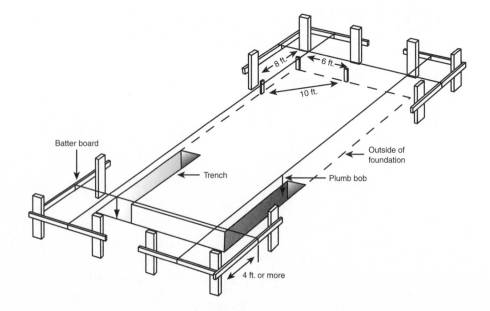

The most common type of foundation is poured reinforced concrete. Because concrete is a slushy mix that dries solid, you'll first need to build framing to contain the concrete while it is being poured from a cement mixer or truck. Fortunately, your foundation sub knows how to lay out and construct the foundation framing. He makes sure that there is a wide base, called a footing, below the foundation wall and that reinforcement bar, or rebar, is placed where the concrete will surround it.

Plans call for the attached garage floor to be concrete slab. Gravel is spread first, rebar is woven where the floor will be, and the concrete truck pours its load over it. Workers will use special tools to flatten and smooth the floor.

Sometime during the site prep, the utility folks have come by to make sure you have electricity to the site. Saws and air compressors will need it. If not available, you might buy some electricity from a neighbor, running a heavy-duty extension cord to your building site. And, uh, make sure that you contact the sanitation rental service to drop off a rental toilet for you and the crew. No sense in upsetting your new neighbors.

On the Level

You may have opted for another type of foundation such as concrete block or even wood. These will be covered in Chapter 18, along with lots more instructions on pouring concrete foundations.

Set: Subfloor and Framing

Your house uses conventional platform frame construction. So, that's what we'll watch in this chapter. Other types of construction will be covered in Chapter 3 as well as throughout Part 4.

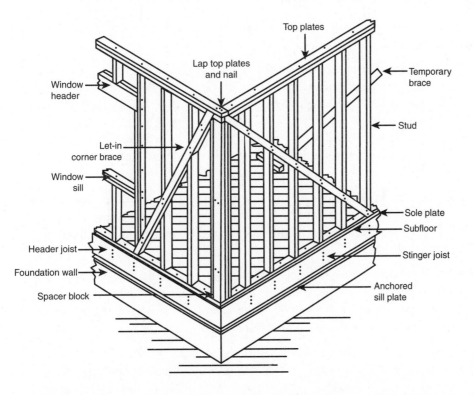

Platform construction details.

Your new house is outlined in concrete. The perimeter is marked, and between the foundation walls stand little evenly spaced pillars of concrete, called piers. There are even a few piers outside of the foundation wall, indicating where your rear deck will eventually go.

The carpenter's crew pulls up and starts moving materials. Then they deftly begin building a frame on top of the foundation. They're adding the sill plate and header. They then install wood (or steel) joists between the outside headers. The framing is starting to look like a large bookshelf that fell on its back.

Typical floor-framing details.

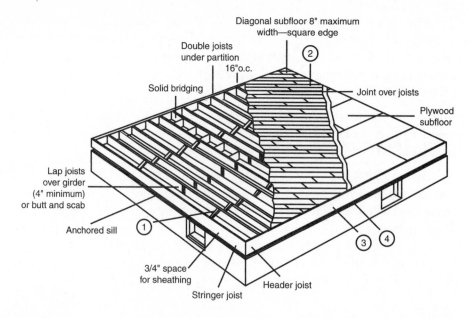

Diagonal subfloor 8" maximum width—square edge

Double joists under partition 16"o.c.

Solid bridging

Joint over joists

Plywood subfloor

Lap joists over girder (4" minimum) or butt and scab

Anchored sill

3/4" space for sheathing

Stringer joist

Header joist

The crew then installs smaller pieces of wood (or steel), called braces, between the joists for support and to keep the joists evenly spaced. You notice the sub periodically pulling out the foundation and framing plans, and then rolling them up and going back to work or calling out a number to the crew. The framing crew also builds a deck frame atop the piers poured by the foundation crew.

Once the framing is in place, the crew starts pulling plywood sheets off the nearby stack and placing them horizontally on the framing. The *subfloor* is starting to take shape.

Building Your Vocab

The **subfloor** consists of plywood or boards installed on foundation joists.

One tradesperson turns on the air compressor on the back of his truck and uses a large pneumatic nail gun to quickly fasten the sheets to the framing. He pushes each into place to get a tight fit before nailing. Another crew-member installs decking on the deck frame. Your house now has a solid platform from which the walls will rise and the residence is built. This is where things seem to happen faster. The full foundation framing and flooring process is covered in detail in Chapter 19.

A framer lays lumber, called plates, atop the subfloor edge. The framer places the two plates, called the sole plate and the top plate, next to each other so that he can simultaneously mark them with a pencil. Depending on the framing plan, he will draw a line across the edge of both plates every 16 inches, and then mark one side with an X. What's this all about?

You soon see, as the framer separates the two plates and places another piece of lumber, called a stud, diagonally between the two plates. He nails through one plate into the top of the stud at 16" intervals. He then repeats the nailing process to attach the other plate to the stud at the same interval. It's starting to look like a smaller version of what the framers did to frame the foundation floor.

All of a sudden, two framers lift one side of the frame until it is standing at the edge of the floor. It's the wall frame! They quickly brace the exterior wall frame so that it won't fall down, and the crew builds and lifts other exterior wall frames. The frames attach to each other at the corners. They build the attached garage walls in the same way.

The crew then starts building smaller interior walls and attaching them to exterior walls. Fortunately, the crew followed the framing plans, and framed openings for doors and windows as well.

If there were a second floor to the house, they would build it in the same manner as the first: subflooring, exterior walls, interior walls. They would leave an opening in the subflooring for the stairs.

It's starting to look like a real house. It becomes more so as a truck pulls up with a load of large pre-built triangles called trusses. A hoist on the truck places them on top of the wall frames. The framing crew swings into action, spacing each truss according to the framing plan. (They could have built roof trusses themselves, but the plan called for "engineered" or pre-built trusses.)

Chapter 20 will fill in the gaps in this quick overview of how walls and roofs are framed.

Go!: Exterior and Barriers

Time to sheath! The framing crew stretches rolls of vapor barrier material over the exterior framing. They then begin cutting and installing exterior sheathing over it. The *sheathing* could be exterior plywood over which they will add masonry or stucco. For this home, they are installing T-1-11, a textured exterior plywood product.

The framers nail the 4' × 8' sheets edge to edge to the wall-frame exterior. They make cuts in the sheathing as openings for windows and doors. The sheets go up quickly and the framing is *really* starting to look like a house!

Building Your Vocab

Sheathing is a structural covering of plywood or boards over wall studs and ceiling rafters.

They pass exterior plywood to the roof where they install it as roof sheathing. They cover the entire roof with the stuff. At places where rooflines join, called valleys, they nail metal strips, called flashing, into place.

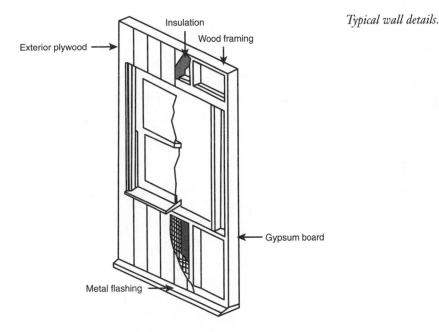

Exterior plywood →

Insulation

Wood framing

→ Gypsum board

Metal flashing →

Typical wall details.

The roofers then arrive to install the roofing material, which makes the roof windproof and waterproof. Roofing material can include sheathing, paper, and asphalt, wood, or tile. Some jobs are complex, but roofing this house is straightforward. Roofing paper is rolled out to seal between the roofing and sheathing. Then the roofers install bundles of roofing shingles, called three-tab, across the lowest edge of the roof, and then as rows toward the top or ridge. They overlap the top of each row with the bottom of the next higher row. Finally, they install special shingles, called ridge cap, at the top to make sure that water runs off to either side of the roof.

Typical construction details.

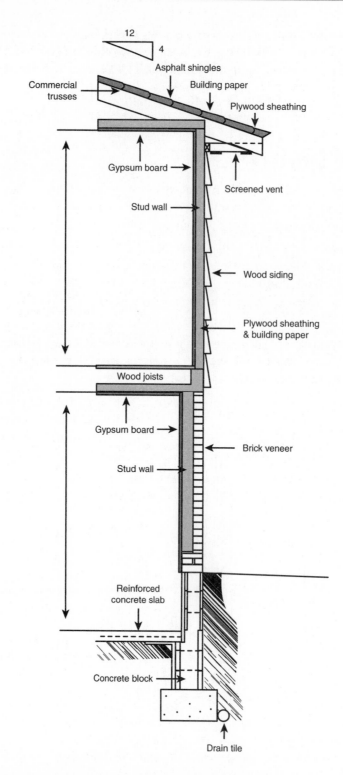

Chapter 21 offers instructions on installing exterior siding and roofing.

Now it's time to install barriers, also known as doors and windows. A flatbed truck pulls up from the building supplier and they offload the wrapped or crated doors and windows.

In turn, the crew places each of the windows in the appropriate opening from the exterior. If everything is working correctly, there will be a small gap between the barrier and the frame. Small wood wedges called shims are placed between the frame and barrier as needed, to make sure it is level and square. They then fasten the windows to the frame. They install doors in about the same manner. Chapter 23 offers instructions on installing doors and windows.

Ka-ching!

The crew leaves the plastic film over the door or window intact to protect it from the painters. The exterior painters will spray everything in sight, including doors and windows if they aren't covered.

Utilities

The mechanical plan (that your contractor-consultant forgot to bring) is vital for installing utilities. It includes one or more plan sheets for electrical, plumbing, gas, heating and air conditioning, and any other utilities or services (cable, home security) that you need.

Here's where the fun begins! As GC, you will try to keep the various subcontractors from bumping into each other while you ensure that the job gets done on time. For example, you may need the plumber to install rough-ins (water lines, sewage lines) in the bathroom at the same time the electrician is trying to pull (install) wiring, the mason wants to look at the floor that will be getting tile, and the cabinetmaker is trying to take measurements. And it's a small bathroom. Fortunately, Chapter 22 will show you how to manage the various jobs using the critical-path method.

Meanwhile, remember that most utility subs have a two-step process: rough-in and finish. For example, the electrician will install the main breaker box and run all of the wires needed throughout the house through the open walls. Once the walls are finished, the electrician will be back to install switches and fixtures. The plumber, too, runs pipe while the walls are open, and adds fixtures once the walls are closed.

More and more homes are also adding other utilities while the walls are open. Things like cable or satellite TV wiring, computer network wiring, intercom or audio system wiring, and home security system wiring are typically roughed in early, and then finished once the walls are closed up.

Interior Finish

A house is not a home until some more things happen. The insulation contractor brings in rolls or sheets of insulating materials, installing them between wall studs, ceiling joists, floor joists, and wherever else needed. The drywall contractor arrives once everything is in the walls. The building materials truck arrives with large sheets of *drywall*.

Deft tradespeople place the large sheets on the ceiling and wall framing, and then nail or screw them in place. They cut and install smaller pieces to make the walls a solid surface. Others cover seams between the sheets with drywall tape. They also plaster over any nails to smooth the surfaces. If in the plans, they will spread or spray a texture material on the walls and/or ceiling. The painters then come in, spraying the walls and ceiling with color.

Building Your Vocab

Drywall, also known as gypsum wallboard, is the interior covering of paper-wrapped plaster.

Windows and doors are then trimmed with decorative material called molding to cover the gaps between the barrier and the frame. The finish carpenters do this. The insulation contractor has already stuffed insulation into these crevices.

Lots more to go. The finish carpenters bring in the pre-built cabinets and install them. They may also install any counter tops, or a mason may add a tile top. As needed, a mason may build or put masonry trim on a fireplace, a porch or doorway, the exterior, or install counter tops. An electrician may connect lighting fixtures and controls. A solar specialist may hook up any solar panels or other products. The utility subs come in once more to do whatever finishing is needed, such as installing faucets, wiring fixtures, and prepping for appliances. Chapter 23 covers all the interior finishing.

Then comes the decorating. Any final painting and trim are added. Curtains are hung. Flooring is installed. Appliances are brought in and installed. Chapter 25 will guide you through decorating your home.

Final Inspections

Ready to move in yet? Not quite. You still need an occupancy permit. All along the way, *building inspectors* have been visiting the site (typically by appointment) and inspecting the house for compliance with safety and health regulations. Who visits and when depends on local building codes and the terms of your construction loan.

Building Your Vocab

The **building inspector** is the local building department employee who inspects the construction of your house to make sure it follows approved plans and building codes for health and safety.

There should be no surprises on the day the inspector comes out for the final inspection. Prior inspections would have pointed out potential problems and code violations, which you and/or your subs have corrected.

Once done, the inspector might sign an occupancy permit right then, or might go back to the office to do so and mail it to you. Or not. If not, the inspector will tell you what the problems are and give you time to fix them before the inspector will sign an occupancy permit.

Technically, the utilities you use (electricity, water, sewer) have been for the construction stage only. You're not supposed to use them for daily living until the occupancy permit is signed.

With the occupancy permit, the lender will probably send out an appraiser to make sure everything was built as planned—and as paid for—before releasing the final funds. Then you can probably convert the higher-interest construction loan into a lower-interest mortgage.

Chapter 24 guides you through the entire occupancy process, including all the paperwork and moving in.

Final Touches: The Yard

It's your home now! (Well, yours and the bank's.)

There's still more you will want to do. The excavator moved soil back up against the foundation, but the building site still looks like a war zone. As a smart construction manager, you've already worked out which of the subs (or all of them) will clean up the building site once work is done. Or you may hire someone else to do so. Or you may do it yourself.

Ka-ching!

Want to actually see a home being built? Watch *Managing Home Construction* as Dean Johnson and Robin Hartl of the popular syndicated program *Hometime* (www.hometime.com, 1-800-227-4888) chronicle the construction of a custom home, from selecting a lot through decorating. Recommended viewing!

Finally, you pull out your comprehensive landscaping plan. It's ambitious, but you get started on some of the primary elements this year, and expect to continue the plan over the next few years until you get the exact yard you want. It will have an expanded deck, fencing, a storage building, and more.

Chapter 26 will give you some ideas and pointers on making your yard what you want it to be.

If you've ever watched a house being built, you know that the description in this chapter is just a snapshot of all the things that must happen in order to build. If you've never built a house before, you now have an idea of what happens. I'll expand on each of the steps later in this book.

So there's really no good reason why you *can't* participate in the construction of your home, from simply being an informed consumer to tackling every job yourself. It's your choice.

The Least You Need to Know

- Conventional homes are built following a standard process and specific plans to make the job come out right.
- If you're financing the construction of your home, working with your lender is an important part of building.
- You can break down each phase of the construction process into tasks and tackle them one at a time.
- Contractors, subcontractors, and tradespeople all know their jobs and move around the building site like ants.
- *You* can participate in building your own home!

Chapter 3

Exploring Your Housing Options

In This Chapter

- ◆ Understanding conventional home construction
- ◆ Learning how nonconventional houses are built
- ◆ Taking a look at kit, log, manufactured, and other homes
- ◆ Should you build a nonconventional house?

Tired? In Chapter 2 you built a virtual platform-construction single-family residence from site prep to moving in. But maybe once you virtually moved in, you found that it really wasn't the kind of home you always wanted. You'd prefer something more spacious (like a timber home), or massive (like a log home), or easier to build (like a kit or manufactured home).

Hey, you've come to the right chapter. This is where we'll consider your options. You'll see how other types of homes are built and why.

So grab your Home Book again and make notes as you go on our Tour of Other Homes. But, before we do that, let's take a look at conventionally constructed homes.

Stick-Built: Conventional Housing Construction

Yes, in Chapter 2 we built a pretty typical, modest home. Nothing too fancy. But maybe you like fancy. Or bigger. Or more environmentally friendly. Or whatever.

Fortunately, so-called conventional construction can handle these options. In most areas, conventional construction is also known as western *platform frame* construction. Why? Because it was developed in the western United States for building what are euphemistically called "ranch-style" homes. Walls are built on the subflooring. From there, each story and the roof are built atop the previous level.

Building Your Vocab

Platform frame construction is a method that uses the foundation, joists, and subfloor as a platform on which the house is built. **Balloon frame** construction is a method that installs studs directly on top of joists. Both use dimensional lumber as their primary building materials, so they are often referred to as "stick-built" houses.

You can also use other types of construction methods. *Balloon frame* construction was popular for two-story structures until the 1950s due to the availability of longer dimensional lumber. Another variation to conventional construction is using steel or aluminum framing members. You can install steel joists under the house. You can include steel or aluminum studs in the walls. You can install aluminum siding on the exterior.

Most residential house plans call for wall studs to be no more than 16 inches from center to center (called on-center, or o.c.). However, some newer technology can allow for wider spacing of studs to save on construction costs. The trick is to make the exterior sheathing take up some of the load. So, you may see this in newer housing plans. But unless you're actually doing the drawings and plans for your new home, you probably don't need to know much about these systems except that they exist.

Lots of Options: Nonconventional Housing

If that's conventional, what's nonconventional? Everything else! Nonconventional building systems, such as post-and-beam and log homes, are especially popular with owner-builders and with specialized contractors. The post-and-beam construction method, shown in the following three photographs, uses vertical posts, rather than rafters, to support beams.

Vertical posts support horizontal beams in post-and-beam construction.

(© Granby Post and Beam Homes)

The exterior of a typical post-and-beam house looks like a conventionally built home.

(© Granby Post and Beam Homes)

The interior of a typical post-and-beam house shows off its wood construction.

(© Granby Post and Beam Homes)

Timber homes are constructed of larger wood components called timbers (duh!). Depending on the construction method, the timbers may be vertically implanted in the soil, with horizontal timbers connecting them, and then walls and windows installed between them.

Pole buildings are similar to timber homes except they use round wood members rather than square ones.

Log homes use horizontal wood members, typically round, though they may be flattened on the top and bottom.

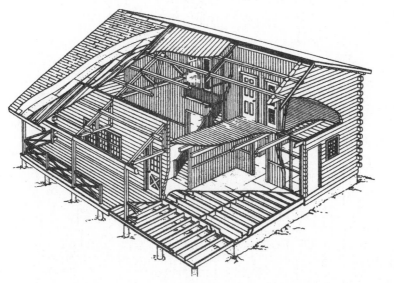

A cross-section of a typical log home.

Masonry houses are typically built from stacked native rocks glued together with mortar. It's lots of work, but can be an inexpensive way of building a house if you have an abundance of rocks nearby. Most masonry houses, though, are conventional houses with a masonry facade.

Adobe homes are built from adobe or clay-and-straw blocks that are then plastered over with an adobe clay mixture. Though beautiful, they require more maintenance than many other systems. Most so-called adobe homes today use a stucco surfacing that needs less maintenance.

Huff and puff and you won't blow a straw-bale house down. They're sturdy. Basically, the walls are stacks of straw bales plastered over with adobe or other material. With a high insulation factor, they are popular for solar homes. Some straw-bale homes use more conventional construction for the interior.

There are other kinds of nonconventional homes. People have built houses from pop bottles, corrugated aluminum, and just about every other material under the sun. But really, there are only two ways to build:

- One level above another
- Hanging from a frame

Conventional construction, log, masonry, adobe, and straw houses are built by stacking something on something else. Walls go on floor framing, roofs go on walls. Timber homes, pole buildings, tents, and other systems typically put up a frame from which walls are hung. That's about it.

Which one is right for you? Most folks select a conventional construction house. They are popular because they are easier to design, to build in place, to find experienced labor for, and to finance, than other types of construction. Even so, you should consider other construction methods and materials than what are called stick-built homes. How can you find out more? Appendix B offers both conventional and nonconventional home construction resources.

To learn about the latest and greatest building ideas, visit local and regional home shows. They will have demonstrations and booths for those who want to participate in building their own home.

Code Red

Don't get stuck with the wrong foundation! With conventional platform construction, you can compensate for a foundation that is a little larger or smaller than the original plans. But there's no fudging with a house that's manufactured to plans somewhere else. So make sure your foundation is built as planned. Be especially sure if you're buying an off-site-built house.

Off-Site Built Homes

Not quite ready to build a straw-bale dome home in your neighborhood? You have other options. You can have some or all of your house built somewhere else and moved to your lot. Really!

Off-site home manufacturers like to say that the most expensive place to build a car is in your driveway. They argue that building cars in factories and transporting them to driveways is actually less expensive. It makes sense. Except that cars are a little more mobile than even mobile homes! In the real world, building some or all of a house off site can save you time and money. Not a lot of money, in most cases, but enough to make off-site-built homes a good option for many home buyers.

So what are your off-site-built-home options? You can build some or all of your house from a kit, have it manufactured in sections and brought in, get it in pre-built panels, or some combination of these options.

Let's look closer at this option that's popular with folks who build their own home.

On the Level

About a century ago, Sears and Roebuck began selling home plans and even kits in their popular catalog. For less than $1,000, your grandparents could get all the materials and plans needed to construct a simple yet functional small (by today's standards) home. The kit even included two small trees to plant in the front yard! Over the next 30 years more than 37,000 of these kit homes were built. Some of them are still serving as functional homes.

Big Erector Sets: Precut Kit Housing

A very popular option with owner-builders is precut kit housing. What's included in the kit depends on whose package you're buying.

For example, one kit manufacturer includes the following:

- Construction plans and instructions
- Floor system
- Wall framing
- Siding and trim
- Roof system
- Windows and doors
- Stairs, if needed
- Garage walls and doors
- Electrical system
- Insulation and sheetrock (drywall)
- Interior fixtures (including bath and kitchen)
- Interior trim and paint
- Lighting fixtures
- Floor coverings
- Cabinets
- Appliances
- Heating and air conditioning

There's even a doorbell included!

Log-home kits are especially popular with do-it-yourselfers. You supply the land and foundation and the log-home manufacturer supplies some or all of everything else.

There are also partial kits available. One of the most popular is a post-and-beam system. The systems include specially milled timbers that fit together like a jigsaw puzzle. Posts are the vertical members, and beams are the horizontal components. Together they become the framing for walls. You can get a full package that includes everything you need, or you can buy a shell package that includes only the wall components.

What's the attraction of a kit house? For owner-builders, it offers a compromise that enables them to build their own home with few or no subcontractors. How much a kit home costs depends on how inclusive the kit is. In general, kits range in price from $25 to $75 per square foot of living space. The lower end is for homes where you supply the foundation and many of the finishing touches. The higher end may include everything from foundation to roof cap. One word of caution: Before ordering any home kit, make sure you know exactly what's included—and what's *not* included!

If you decide on building a precut kit home, you haven't wasted your money on this book. The financing, foundation, and construction process are exactly the same as for conventional construction.

This cedar log home was built from a kit.

On the Level

My popular book *Building a Log Home from Scratch or Kit, Second Edition* (see Appendix B), is still around in used books stores, or look for it at online bookstores such as www.amazon.com and www.MulliganBooks.com. It describes the log-home construction processes for both log-prep and kit methods, and features more than 300 illustrations.

One from Column A: Panelized Housing

"Yes, sir. One house to go! How many walls do you want with that?" You can purchase full walls complete with framing, sheathing, and even windows and doors for your new house. Depending on the manufacturer, you may be able to select from a few wall designs or your own plans.

Frankly, panelized housing isn't as popular as one might think from the concept. The reality is that most walls still need to be finished on the inside, and the final costs aren't much below conventional stick-built houses. Panelized walls are more often used in timber-frame homes, actually hanging the precut walls between timbers.

By the Slice: Modular Housing

One step up from panelized housing is modular housing. What's the difference? What's been done for you.

Modular housing offers you a home pre-constructed in modules or sections. The most popular type of modular housing is what's called manufactured housing, so called because these modular units are built in the same assembly-line process as cars are manufactured. Manufactured housing is distantly related to what used to be called mobile homes. It's an entire house broken up into two or three side-by-side modules that are towed to your building site and placed on a foundation, and then connected together.

The foundation for a manufactured home.

A modular house is built in a factory, typically following the same building codes as for conventional housing. It begins with a steel frame on wheels. One group of workers adds the flooring, which is then rolled to the next station where another group builds and erects walls including electrical and plumbing. At the next station, they finish the interior, install cabinets, and attach fixtures.

Modular housing is growing in popularity. More than 20 percent of all new single-family residences in the United States are manufactured homes. It may seem like cookie-cutter housing, but some manufacturers can extensively modify stock plans to make a semi-custom home. You'll even see modular construction in attractive two-story houses, apartments, motels, and commercial buildings. Rooms are built in a factory, and then hoisted into place like building blocks. They are very cost-effective.

> **Code Red**
>
> Make sure any manufactured home you purchase is built to HUD Code. It's also known as the Federal Manufactured Home Construction and Safety Standards, the only federally regulated national building code. The code is administered by the U.S. Department of Housing and Urban Development (HUD).

I recently helped my mom and dad select, purchase, place, and move into a manufactured home. We were able to modify a stock design, flipping the entire plan to fit the lot, turning a family room into an enclosed porch, and adding a garage and deck. No one driving by would guess that most of the construction work was done 100 miles away in a factory.

One advantage of modular housing is that you can move in faster. With the right help, you can be in a modular house within six to eight weeks. But be aware that glitches do occur. In my case, it took nine *months* to get occupancy of the manufactured home for my parents. Everything went wrong: a dishonest dealer, bureaucratic red tape, the land variance, and an inept setup crew. Without constant pestering and withholding of funds, the move-in would have taken more than a year. Fortunately, though, the experience added to my education.

This manufactured unit (one third of the finished house) is trucked to the site and placed on the foundation with a large crane.

Remember, modulars are *not* mobile homes. Once they're set on a conventional foundation and the axles and wheels are removed, they cannot be moved any more easily than a conventional house. Why the distinction? Because in many places, mobile homes are personal property, and manufactured homes are considered real estate.

Selecting the *Best* Housing Option

Should you consider building a nonconventional or factory-built house as your home? Yes, you should consider it. Consider all of your options. Spend some time looking at log-home kits, manufactured houses, panelized housing, and any other options you can think of. Talk with lenders, other owners, dealers, contractors, and anyone else who can give you some insight. Of course, remember that each resource has a bias. Everyone wants to sell you something. The best resources are current owners of the types of housing you're considering.

Financing may be an issue in your selection. It is typically easier (and cheaper) to get financing on a conventional house than on a nonconventional one. But you still have lots of options. Most manufactured and kit-home dealers can advise you on finding friendly lenders. Or you may be better off asking a conventional lender if they can finance your unconventional house. Some can; some can't. If so, their rates may be lower than working with the dealer's lender. Chapter 10 will show you how to get financing for your conventional or nonconventional house.

Why should you consider other types of housing? Because you may get more of what you want for the same, or even less, money. And that's why you're reading this book.

The Least You Need to Know

- In many locations and for most housing needs, conventional or stick-built housing is the best option.
- Nonconventional housing can be the best option for unique styles or lower cost but may require special knowledge or experience to build.
- Kit (or precut) homes have been popular for a century because they lower costs and make construction easier.
- Building your own home, conventional or not, offers an opportunity to express your individuality.
- Most of the information in this book about conventional construction will apply to other types of housing you choose.

Chapter 4

Who Wants to Be a Thousandaire?: Home Savings

In This Chapter

- Estimating the approximate cost of your home
- Building smarter: strategies that can save you lots of money
- Saving on contractors and suppliers
- Using a computer and software programs to lower costs
- The importance of insuring against loss

Chapter 1 berated some of the silly claims that are made about how much you can save by building your own home. Yes, you *can* save many thousands of dollars by tackling some or all of the job yourself. However, in the *real* world, most folks get more house for their money by building their own home. So can you.

This chapter offers dozens of proven ideas for saving money *and* getting more for each dollar you spend on your house. Besides some smart design tips, it shows you how to estimate costs, save money on contractors and suppliers, and use a computer to design and build your house. It also takes a look at insurance.

So grab your Home Book again and start making notes on ways you can get more for each dollar you spend building your new home.

What's in Your Wallet?: Estimating Costs

Can you afford to build the house you want? In Chapter 9 I'll help you get a comprehensive estimate of your construction costs. For now, though, let's make a rough estimate.

First, understand that whatever house you build will probably cost as much or more than that of your neighbors—unless you plan to do it all yourself. Based on this fact, let's look at your neighbors' houses.

As mentioned in Chapter 1, the typical house built today is about 2,300 square feet in size, and sells for around $225,000 depending on where you live. The actual price in your area may be half of that or 10 times that amount.

So, how can you get a good rule-of-thumb for your area? Of course, you can talk with contractors who will throw out numbers like $100 or $200 per *square foot* (s.f.). That's a wide range. What does it include? House only? House and land? Landscaping, too?

It's smarter for you to get your own local building costs. Here's how. Talk with a local real estate agent about new construction. It's best if you can compare apples to apples, so ask about new homes and lots in the area you want to build. The agent should be able to tell you the lot cost, the asking price, and the square-foot size of a few comparable homes. You can figure things from there.

Building Your Vocab

A **square foot** (s.f.) is an area one foot (12 inches) by one foot, or 144 square inches (s.i.). A **square** is 100 square feet.

Code Red

Be careful when considering other people's square-foot cost estimates. The number may be for a house of lesser or greater quality than the one you want. More important, they may have factored in the land cost, dividing the home price by the square footage. Or, they may be attempting to prepare you for a high bid.

For example, a $250,000 new home (comparable to what you want) on a lot costing $60,000 means the house is valued at $190,000. It's a 2,200-s.f. home, so you calculate the value at just over $86/s.f. You do calculations on a few other comparable new homes and come up with a local rule of thumb of $90/s.f.

You now have a base from which to guesstimate the approximate cost of your new house. Then you can factor in the following:

- If you want a nicer view lot, add to the land costs.
- If you will be your own general contractor, reduce the total by 10 to 20 percent.
- If you will be a subcontractor, reduce the total by the sub's fees.
- If you want to enhance the design, tack on the additional costs or get a square-footage cost for better quality homes.
- If you will be building the house yourself, reduce the total by about half if you're not paying yourself.
- If you already own your land, factor that in to your total construction costs.

Money-Saving Strategies

Where to start?!

The first place to begin saving money is on the design of your house. A smart design can easily save thousands of dollars on construction. Conversely, an unnecessarily complex design can really add to your costs. Let's take a closer look at how it all adds up.

Add It Later

Most folks who build their own home try to put every option into it—and a few go broke doing so. If you're trying to keep costs down, consider what you can add later. For example, if you want a central vacuum system, make sure the design includes it. Install the vacuum lines in the walls, but don't purchase and install the vacuum unit until you're into the home. The same goes for wallcovering. Paint for now, and plan to add wallcovering later. It will keep your building costs down and make sure you qualify for that construction loan.

Depending on how tight your budget is, you may choose not to include appliances in your construction loan, rather using what you have for a little while longer, and buying new ones as needed.

You can even add a deck and landscaping later, once you're in your house. Some folks opt to take out a second mortgage in a year or two, to add things they couldn't quite afford when the house was built. If this is your plan, include those ideas in your building plan as needed.

Cut Corners

The cheapest house you can build is square. Only four corners. The more corners you add, the higher the cost. Certainly you want more than just four walls in your house, but remember that each corner you add costs more than a straight wall. And rounded sections, such as bay windows and arched doorways, cost more than conventional walls and doorways—typically about 100 to 200 percent more.

Not to worry. These are just real-world building facts. Make sure you get value for each of these additions to your plan. Keep them to a minimum.

Don't Change Your Mind

Changing your house plan any time after it is drawn costs money. That's a fact. How much it costs depends on what's involved in making the change. If you're simply changing out one window style for another of the same dimension, the cost difference will be minimal. If you want to replace a window with a sliding glass door that opens on to a new porch, you're talking more bucks.

Again, don't worry about it. Just be sure you make all or most of your changes on paper where they are cheaper. In fact, that's one of the purposes of your Home Book. It's a combination scratch pad, list holder, address book, and design-idea notebook. Make sure that the final house plans incorporate those ideas you feel are important to your new home.

Forego the Basement

Basements can be practical additions to a house. Or not. If the basement in your plans doesn't have valuable function, get rid of it. In some parts of the United States, the basement is where you live during tornado season, making it necessary. Or a basement may be required due to the building site slope. So, design the basement for multiple functions rather than just as a storage area. The same goes for each area of your next home: make sure it has more than one practical function.

Build Up, Not Out

The wider your house is, the more it costs. That makes sense, right? You can save money on your house by building up. A second story costs less to build than doubling the size of the first story. Why? Because you don't need twice as much foundation. The first story becomes the foundation for the second story. Of course, the foundation for a two-story house needs to be designed for the extra weight and stress, but it's still cheaper than twice as much foundation. A higher home can also put more house on a smaller lot, saving on land costs.

Keep the Driveway Short

Keep your driveway as short as you can. Yes, it's nice to be as far off the road as possible, but remember that each foot away from the road adds to the cost of your house.

Certainly, one of the costs is for the pavement needed to get from the road to your garage. But there's more. Utility hook-ups (unless you are on a well and septic system) are at the road and must be piped back to your house. This can add hundreds and even thousands to your costs. Adding an extra pole on your property for a long electricity and cable run can cost a couple thousand dollars or more.

For privacy, instead of moving away from the road, plan for privacy fencing or landscaping. It's less expensive.

Keep Down the Amount of Trim

Even the most plain house is enhanced by what's called gingerbread, or ornate trim. A few hundred dollars of trim can really add to the look of a home and increase its value, but it adds to the cost of the house because it takes more time to install. Trim can mean anything from decorative roof edging to impressive landscaping to an especially beautiful front door.

Even More Ways to Save Money

Need some more designing ways you can save on the construction of your house? Try these ideas:

◆ Consider engineered floor trusses to span greater distances and reduce material costs.

◆ If building code allows, don't install bridging between joists.

◆ Try to match door and window openings, as well as interior wall intersections, with the location of exterior studs.

◆ Keep stairways simple.

◆ Consider 24" o.c. framing with members above each other (rafters above studs above joists) if allowed by code.

◆ Centralize utilities with back-to-back plumbing and bundling electrical runs.

◆ Keep plumbing on interior walls where less insulation is required.

◆ If code allows, use plastic plumbing pipes instead of metal.

◆ Use smaller studs, such as 2 × 3s, for interior walls that don't bear weight.

◆ Eliminate wall blocks that aren't required by the building code.

◆ Buy prefabricated shower and tub enclosures.

◆ Don't overbuild. Keep your house costs and quality similar to that of your neighbors' houses to retain value and to better recoup your investment.

> **Ka-ching!**
>
> Concrete slabs typically cost less than concrete perimeter foundations. Depending on local building codes, site soil, and your home's design, you may be able to save with a slab foundation.

> **Code Red**
>
> Be aware that some general contractors also make money by requiring a fee from subcontractors for getting the job. Other unscrupulous GCs and subs will install lower-cost materials than what is specified and paid for, pocketing the difference.

Saving on Contractors

Earlier in this chapter and in Chapter 1, I mentioned that the general contractor's fee is 10 to 20 percent. Actually, it can be even more!

A general contractor typically gets a discount on building materials from suppliers, some or none of which may get passed to the owner. When materials run $100,000, this discount can be significant and add to the GC's total income. Certainly, the GC earns some of it by coordinating the purchase and delivery of the materials as needed by the subcontractors, but a smart owner-builder can do this.

For many contractors, this discount is their profit margin. The contractor won't work except for profit, but it's coming out of your pocket, so you don't want it to be excessive. Or maybe you want to have it all yourself, using your savings to enhance your home. Makes sense. Chapter 13 thoroughly covers what it takes to be your own building contractor.

How will you know how much contractors and subs will charge you for building your house? Once you have a specific design down on paper (like the plans you saw in Chapter 2), you'll get contractors to bid on construction. We will cover how that's done in Chapter 12.

Saving on Suppliers

You'll probably need more help than just contractors. If so, you should know how much each of these services charge and what they do. You may decide to do it yourself and save the money. (Chapter 15 offers ideas on saving money as you hire suppliers.)

Architects

An architect does much more than just draw up house plans. That's a drafting service. The architect helps you design the house to fit your needs and budget.

How does the architect get paid? Like lawyers, some are on an hourly rate, but most are hired for a percentage of the project value. Fortunately, they typically charge less than lawyers. A fee between 8 and 15 percent is most common. That's $20,000 to $37,500 on a $250,000 house.

It's really tempting to not use an architect and to save the fee. What are your options?

- Design the house yourself, and use a drafting service to draw up all the plans.
- Buy a set of stock plans and follow them.
- Buy a set of stock plans and have an architect or drafting service revise them with your changes.
- Negotiate a lower fee (or an hourly rate) with the architect for reduced services.
- Hire a qualified architecture student to revise your stock plans or develop new ones for you.

Real Estate Agents

A real estate agent can be very useful in your search for buildable property. He or she can also help you calculate local building costs as suggested earlier in this chapter.

How much do real estate agents get paid? They typically get a 5 to 10 percent commission on whatever they sell. The low end is for new and preexisting houses. The high end is for vacant land.

There's lots of room for negotiation on commissions, especially if you're dealing with the broker rather than an agent. The broker is the boss, the one who must have a special license and more bonding than the agent (actually, a salesperson) who works for the broker. Depending on the marketplace, you may be able to negotiate a price reduction and lower commission with the landowner and the broker.

 Ka-ching!

If you buy the property directly from the owner, you can save the entire commission. Just make sure you know what you're doing or have legal advice to make the transaction go smoothly. Consider hiring a real estate attorney on an hourly basis to draw up the papers.

Materials Suppliers

Somewhere around half of the cost of your house will go toward materials—everything from concrete to lumber to plumbing fixtures and roofing. How can you save money on the cost of these needed materials? Just like most things, shopping around can make the difference.

Start with your major materials: lumber and hardware. A little local research should tell you where the building contractors buy their materials. Then ask to speak with the contracting department and tell them you're an owner-builder shopping for a materials supplier. Ask about discounts, credit lines, references, materials availability, and delivery (including fees).

Chapter 15 will show you exactly how to hire suppliers for your home construction project. For now, you want to weed out suppliers who don't want to work with owner-builders and whom you don't want to work with. Once you have your building plans, you'll approach them again for bids and take the best package.

Other Services

You can save money on other construction services by being a smart consumer and hiring people with the skills you don't have.

One service that many owner-builders use is a permit service. As you will learn in Chapter 11, there is a whole ream of paperwork that has to be developed and processed before anyone pounds the first nail. A permit service helps you through the building-permit process, from determining what is needed to preparing and filing plans, representing you at planning meetings, and making sure the bureaucratic process goes as smoothly as possible.

Most permit services have an inside track at the local building department, a track as good as or better than most local builders. That's what you want. You'll pay for it, yes, but it is a good option if you are your own GC.

You can also hire a building advisor. It's usually a building contractor or other experienced person whom you can call on as needed throughout the planning and building process.

Where do you find these other services? Most are in larger cities and towns. A few you can even hire and they will advise you by e-mail. Ask other owner-builders, check area telephone books, and watch for ads in trade publications. Also take a look at the resources listed in Appendix B.

Computers and Software Programs

Depending on who you ask, computers are the greatest or the worst invention in the last 100 years. As a computer user since 1984, I'd say that both are true. Actually, computers are simply tools. They do one thing *very* fast: They process information at lightning speed. Software programs tell the computer what and how to process the information.

Most building contractors use computers and software programs to help them manage projects and bids. So can you. In fact, software can give you the edge you need to manage your building project. It can help you with the designing, estimating, scheduling, paperwork, communications, and other vital tasks. Construction software can make sense of the process—or make a mess of it. Where can you find out more about construction software? In Appendix B, of course. I'm not going to recommend specific software programs, because they keep changing; yesterday's leader may become tomorrow's dust-collector. I will recommend that if you don't already have a general understanding of how to use a computer, you get yourself taught. Take a class, read a *Complete Idiot's Guide* or two, hire a tutor, or ask a computer-literate friend to help you. Certainly, you can build your own house without a computer—millions have done so—but it can make the job easier and possibly save you some money.

Add It Up: Estimating Software

Computers are at their best when they are doing math. Great! That's exactly what you want the computer to do as you try to figure out how much your version of the Taj Mahal is going to cost.

Job estimating software can not only crunch numbers, it can tell you which numbers to crunch. For example, one popular estimating program called EstimatorPro includes categories and fields for each of the primary tasks and materials needed in typical construction. It even includes proven ratios and rules of thumb to help with the estimating process.

Estimating software isn't cheap. If you'd rather do it yourself, use one of the popular spreadsheet programs that can calculate data you give it. In fact, most estimating software is based on a spreadsheet with most of the data requirements already filled in.

Write a Check: Money Software

There are many programs out there to help you keep track of your money, such as Intuit Quicken or Microsoft Money. They can be as simple as checkbook software that is a computerized check register, or they can be an entire accounting system with tax and payroll modules to help you account for every penny.

The disadvantage to money software is that it takes time. The advantage is that it saves time. The *real* advantage, however, is the reports that you can easily get from such programs. Want to know how much each of the subs has received thus far? Just a second. Need to know what bills are planned for payment at the next draw? Just a second. Have to reconcile your bank statement to make sure there are no errors? Yep, just a second.

Ka-ching!

Freeware is free software. Shareware is software you try out and, if you like it, you pay the programmer. You can find free and low-cost software at sites such as www.freeware.com and www.shareware.com.

Browsing the Internet

You can find just about anything on the Internet. For example, I recently finished replacing about 600 square feet of flooring in our house. I've been able to compare various products, get some installation tips, request a sample, speak with other users, and even order and pay for the flooring online. Amazing!

To do such things, your computer will need what's called a browser program. Once you have access to the Internet (through a modem and an Internet service provider, or ISP), the browser will be your window to the world. You can look things up using a search engine such as www.google.com.

Paying for things online is relatively easy and, more important, secure. Like anything, it takes some common sense to not give your valuable credit card information out without security. But I've purchased many thousands of dollars of products and services online without a single problem (knock on laminated wood).

An e-mail program will let you correspond with suppliers, manufacturers, and other owner-builders through their e-mail addresses. E-mail programs can also give you access to user groups. These are groups of folks with similar interests who correspond with each other electronically. Anyone in the group can post a message that everyone in the group sees. You can then respond to the group, to the individual, or both. It's a great way to get other opinions. (It's also a pretty good way to waste time.)

Are You Insured?

Let's talk about one more thing before we leave this chapter in the sawdust: insurance. The building trade is just about the most dangerous trade this side of hand-feeding sharks. Otherwise smart and careful people fall off roofs, get hit in the head by lumber, trip on debris, cut off fingers, and do other things to make their lives interesting. And your life is expensive! Yep, you're liable for everything that happens on your building site. That's right, *everything*. Even some kids playing around the site on the weekend; it comes under *your* liability. You don't want those kids playing at your building site to wind up owning your new house!

If you have a general contractor and/or subcontractors, make sure they have signed something to release you of liability. That is, *their* insurance covers what they do. If you're not using a GC or subs—and even if you are—talk with an insurance agent about covering your building site and all the bad things that can happen there. Ask about a "course of construction" policy. Make sure you or your contractors have adequate worker's compensation insurance in case one of their employees is hurt at your job site.

The Least You Need to Know

- ◆ You can estimate what your home will cost you by calculating a local square-foot rate.
- ◆ A smart home design can save thousands of dollars on construction. Consider such money-saving strategies as adding features later, building up instead of out, and keeping trim to a minimum.
- ◆ Be a smart consumer to save on architect fees, real estate agent fees, and materials suppliers.
- ◆ Put your computer to work to save time and money by using construction software programs, or by browsing the Internet, to compare products and speak with other users.
- ◆ You are responsible for everything that happens on your building site; make sure you have insurance to protect yourself against loss.

Part 2

Designing and Paying for Your Home

See? You can build your own home! So it's time to get a little more serious. Specifically, it's time to begin designing, locating, and figuring out how you're going to fund this dream. Part 2 has you covered!

In this part, you'll learn how to design a home that fits your needs rather than those of the builder, how to find the perfect area and building site for your dream home, and how to estimate construction costs so that you squeeze every drop of value from every dollar you spend. You'll also learn how spending a few extra bucks now will save energy costs in the long term. And I'll tell you how to find a smart lender who will invest with you in your dream home at the lowest rates.

Get ready to walk the path to your new home!

Chapter 5

How Many Bedrooms?: Designing Your Own Home

In This Chapter

- ◆ Getting design ideas and selecting the best
- ◆ Figuring out what you can afford
- ◆ Making the right compromises
- ◆ Hiring the best designer for your house
- ◆ Producing your own home plans

Now comes the fun part! You get to play around with various ideas until you come up with the best design for the home you're going to build. That's what this chapter covers: designing your own home. Yes, it will help you come up with new ideas and try them out. But just as important, it will take a practical look at your design to make sure it's really what you want and what you can afford.

The chapter will also help you decide whether you need an architect or a plan service, or whether you will develop your own plan set. As before, make sure your Home Book is handy. Let's go have some fun!

Getting Ideas

Maybe you already have some good ideas for what you want in your next house. Or not. In either case, keep an open mind as you consider how best to design your home.

For example, you may find that arched doorways from a cottage design would really enhance the feel of your ranch-style residence. Or maybe you'll borrow ideas from a log home for your shingle house. So keep an open mind.

Get Lost!

There are numerous architectural styles available to you. And each has components you may want in your home. The best place to start looking for ideas is in the neighborhood in which you plan to build. If you haven't decided yet, get ideas from throughout your favorite housing area.

Take a drive through these areas, Home Book and camera or video in hand. Make sure you also take anyone who will participate in the design decision.

Most of today's new homes aren't pure anything. They may be based on one style, such as a Victorian, but probably include ideas from many other styles. Even plans in the stock plan books illustrate hybrid architecture designs. So feel free to mix and match.

Make a New Plan, Stan

Speaking of plan books, you'll find literally thousands of house design ideas on magazine racks and in book stores. The plan books and magazines often focus on a style or size of house, but others will offer a sampler of traditional, farmhouse, colonial, craftsman, contemporary, and other popular styles.

What are they selling? Plans. You can purchase full sets of plans for these houses from plan services. We'll discuss what's included later in this chapter. For now, buy a few of these plan books and magazines for ideas. You may even find your dream home—or a reasonable facsimile.

Don't Leave Home

You can also get some good design ideas without leaving home. You can visit other homes in your mind. Think back on the home(s) in which you grew up. What did you like or not like in these residences, design-wise?

Also visit your friends' and relatives' homes, in your mind or for real. Consider the various design features and glean from the best.

Watching TV can help you see various house plans used in sitcoms, decorating shows, and other entertainment. Football games don't count.

Say, Home Book!

Let's talk more about your Home Book. By now you've made a few notes on initial thoughts and ideas for your house. Great! Don't stop there. Start asking those in your living group what they would like to see in the house.

Kids might say "swimming pool!" Or they might say, "My own room." Whatever their response, ask them to elaborate. They might realize that a swimming pool isn't that important, but a separate room or at least a larger shared bedroom is.

Assessing Needs and Wants

Okay, time to start closing your mind. That is, you now need to begin narrowing down the search. Get more specific. The first step is estimating your space requirements. How much housing space do you need?

Before shooting a list of questions at you, let me remind you that space costs money. If you figured that new construction in your area costs $100 per square foot, a 10' × 12' room costs $12,000, and a 15' × 16' room is double that amount. In addition, a bigger house can mean a bigger building site, which equals more money.

It's smart to design backwards from the money. If your budget says you can afford a $250,000 house and, in your area, that means 2,200 square feet, make sure your design is that size or less.

On the Level

A square foot is an area 1 foot long by 1 foot wide. That's easy. However, calculating square feet (abbreviated sq. ft.) of a living area can be more difficult. It's easy to calculate the square footage of a house that's 60 by 40 feet (2,400 sq. ft.), but what do you do if your home plan varies in width and length? You break up the space into squares and rectangles, calculate the square footage of each area then add them all up. For initial planning, an approximate square footage is sufficient. However, as you start designing your final plans you'll want exact measurements and size.

Okay, now the questions:

- Do you prefer a formal or a casual house—or a mixture?
- How many bedrooms do you need?
- How big does the master bedroom need to be? And does it need walk-in closet and separate bathroom?
- How about the other bedrooms? How long will children be at home? Do you need to set aside a guest room, or can it also be the den, music room, or other special-use room?
- How many bathrooms do you need? Closest to which rooms? Tub, shower, both, or neither?
- Do you prefer a single large living/great room for unity or a smaller living and family room for separation?
- What do you want in your kitchen? Are you microwave mealers or weekend chefs?
- Is a separate laundry room important, and should it be near the living area (for convenience) or in the basement or garage (for space)?
- Are TV and games important enough to your lifestyle for a separate media room?
- How many cars will need to be garaged? Attached or detached? How much storage space do you need so you can park the cars inside?
- What do you like the neighbors to see and to not see? What style and landscaping do you prefer for the front and rear of your new house?

In your spare time, consider lifestyle questions that will help you design the best home for you and your living group.

- How much time do you spend at home? In what part of the house? Doing what?
- Where would you *like* to spend your time at your new home?
- Where do visitors like to congregate? Kitchen? Living room by the TV? On a back deck?
- Where would you *like* to have visitors congregate in your new home?
- What would you like to add to your house to make it special for you and your living group? A game room? A commercial kitchen? A big shop off the garage? A kid's outdoor play area?
- What about lighting? Natural, fluorescent, incandescent or all of the above?
- Any preferences for heating systems?

Write the answers to these questions in your Home Book, subject to change.

Visualize the Plan

It may help in your designing process to draw circles on a piece of paper, large circles for large rooms and small circles for smaller ones. It's the first step in visualizing your home plan.

Next, draw connecting lines between those rooms that should be near each other. For example, connect each bedroom with the bathroom that will serve it. Connect the garage to the kitchen or to a utility or storage room. Connect the stairway, if any, to the rooms it serves. Do the same with the entryway.

On the Level

Make sure you keep rooms, closets, and hallways in scale. If in doubt, measure these design elements in your current abode as references.

This little exercise can help you begin visualizing the internal structure and flow of your new house. It will also help you in talking with an architect or reviewing plans in a plan book. It's a lot cheaper to move rooms around on a piece of paper than on a blueprint—or a built house. I'll show you how to take your doodling one step further in the upcoming section on "Making Rough Plans."

Living Inside the Box

Because houses are typically built of squares and rectangles, the next step is to start designing your house as such. That doesn't mean every room has to be a perfect square or rectangle, but it should at least start out that way while you shuffle them around.

One way of making this job easier is to create paper cutouts of each room and move them around. You don't have to know the exact dimensions, but approximate size will help you in placement.

Alternatively, you can use a home design software program (see Chapter 4) to block out and resize your rough plans.

Precut and reusable symbols can help you plan your home.

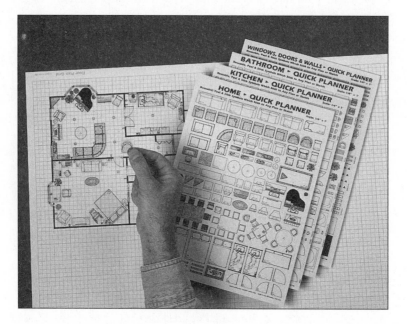

This is an important step whether you're going to design the house yourself, hire an architect, or buy a stock set of plans. Knowing more of what you want will help you eliminate what you don't.

Who Did Your Decorating?

Even though you haven't decided on a final house design yet, consider hiring an interior decorator—or at least looking through decorating ideas—with your Home Book in hand. Why? Because your decorating ideas can suggest structural changes in your house.

For example, wanting a small breakfast nook may suggest a nearby window treatment. Or selecting a large sectional couch for the living room may suggest the room's layout. Or deciding on long drapes in the master bedroom may suggest a higher ceiling. Otherwise, once moved into the house your great decorating ideas may not fit.

> **Ka-ching!**
>
> Interior designers are paid by the hour or as a percentage of the furnishings you purchase. In fact, an interior design store may *not* charge you for decorating consulting. That's because the designer works on a commission of the furnishings sold.

Life Is Compromise

You know approximately what you want. And you know about what you can afford. But the two don't seem to match. Welcome to the *real* world of building your own home!

It's about compromise. You really didn't think you were going to be able to have an indoor swimming pool, did you? Why not compromise with an indoor spa or a bathtub with water jets? Or build near a health club.

Actually, by building some or all of your home you can probably afford more than if someone else builds it for you. A nice incentive to doing more of it yourself.

Compromise isn't easy. The best place to start is by reviewing the things you want in your next house and marking them as one of the following:

◆ Need

◆ Want

◆ Wish

For example, you may decide that you want a full master bedroom, but you only need a standard bedroom with a bath nearby. Or you may need a full master bedroom and wish it had a spa built in. Or you may need a full master bedroom with spa and want a spa large enough for block parties. You get the picture.

So review the design features listed for your home identifying their priority. Again, building some or all of it yourself may put some of the "want" list in your "can do" column.

> **Ka-ching!**
>
> Consider semi-custom. That is, if standard construction is $125/s.f. in your area and custom construction is double that, consider budgeting $150/s.f. and add your favorite amenities to a stock home design.

Gimme Some Space, Man!

As you're designing your next house, think of it as spaces as well as rooms. What kind of space?

◆ Communal (family, visitors)

◆ Work (home office, garage shop)

◆ Private (bedrooms, hobby)

◆ Service (kitchen, baths, laundry)

◆ Circulation (entryway, hallways)

The advantage to looking at your house as clumps of spaces is that it can help you place them by use. For example, if possible centrally locate service rooms such as kitchen, baths, and laundry. Make sure that private areas are away from work areas. Plan circulation spaces to fit the adjoining needs.

Uh, Anything Else?

Yes. First, a sermon. Design is the most important step in the construction of your house. A carefully thought-out design that considers all uses now and into the future will make it a more enjoyable home. Smart designing can make life easier.

As important, a good design that considers the construction process can save you lots of money. Well-drawn plans tell your contractor and subs *exactly* what you want and saves everyone time (read: money).

So the moral of this sermon is: Design for success. Many owner-builders work on their design for a year or more before actually starting construction.

Okay. You didn't fall asleep during the sermon, so here are some additional tips that can help you fall in love with your home even before you build it:

- Make it *your* house. Design features and amenities into it that reflect your personality, interests, and life. Otherwise, go buy someone else's house.
- Make it fit the location. Don't try to build a log home in a small-lot subdivision nor a mansion in the woods. Instead, bring a home to a location where both look at peace with each other.
- Fit the house to the location. Place doors and windows where they can take advantage of the sun, trees, nearby water, and other natural features.
- Design from the inside out. Use the design principles in this chapter to picture the home you want to live in before you plan the home that others will see.
- Extend the inside outward. Add decks and porches as outdoor rooms at lower cost.
- Remember that rooms are three-dimensional. Plan to use higher ceilings where spaciousness is needed.
- Respect the land. Build with the least destruction to the property, the view, and the environment. Thank you.
- Be realistic.

Don't Stop Thinkin' About Tomorrow

Okay, you're *never* going to move from your new house. Ever! If land-use laws permit they're going to bury you in the backyard.

Well, someone's going to sell it someday. (So you may get an unmarked grave.) So what?

So remember to design your home with an eye on resale. That is, don't try to build a seven-bedroom house unless you really need seven bedrooms. Not much call for them anymore. Also, make sure you follow all building codes because you probably can't resell it if you haven't.

The point here is to consider that unique homes have a limited market. That means they don't sell as fast (or at all) or for as much money as custom or semi-custom standard housing.

Making Rough Plans

You really don't need an architect to tell you what you like. You can make up your own building plans—or save on the cost of architectural plans—by doing all or some of the planning yourself. All you need to start is a sketch of the room functions and sizes you want. Let's give it a try.

Sketching by Function

It doesn't matter whether you're an artist or not, you can start drawing plans for your new home. That's because you'll start with circles. Simply draw circles on a sheet of paper and include in them related functions. For example, one circle might include the functions of a family room. Another might be a game room. Another circle from your function list includes a bathroom. Another is a laundry room. Lots of circles, each with a specific function.

Next, draw lines between these circles to show any relationships. That is, a bedroom needs a nearby bathroom. A kitchen serves the family room or recreation room. A home theater doesn't require other rooms so it stands alone in a circle. These are rooms that need each other.

Finally, use a different color pencil to draw lines between circles of similar use. That is, a daytime office space also can serve as a home entertainment space during the evening and on weekends. A children's play space can serve double duty as a hobby space. Remember, these aren't rooms yet, just functional spaces.

If you have a computer, you can use one of various software programs to do these functional sketches (covered later in this chapter). Because these are just circles and lines, simple drawing programs found on most computers will serve the purpose for now. You're attempting to help visualize the relationship of functional spaces. Use whatever tool, pencil or computer, that makes it easy.

Sizing Functions

The next task in planning is to estimate how much space you'll need for each of these functional areas or circles. For example, measuring an existing bedroom in your home you may determine that a new master bedroom will require 180 sq. ft. to be practical. Or you may need at least an 8' × 10' (80 sq. ft.) space for your home office. Write these numbers down in the appropriate circles on your functional plan.

Now total it all up. Your functional plan and wish list may include seven functional areas that require 800 sq. ft. or more of space. If your design goal is a functional home up to 3,000 sq. ft. the final decisions are easier. If, however, the functional areas you want don't match the available space, you have some more planning to do. Fortunately, all the circles and lines you just drew will make trimming much easier.

In the real world, you may come up with two or three different functional plans. That's okay. The process is about considering all options and requirements, then whittling it down to the best. Remember: Making changes to your plan on paper is *much* cheaper than making construction changes!

Architects Do It by Design

Do you need an architect to help you design your new home? Probably. The real question is: Do you need to hire an individual architect to plan a one-off house or can you buy stock plans designed by an architect?

Unequivocally, that depends.

If you have the budget and your home isn't cookie-cutter, certainly consider hiring a professional architect. The American Institute of Architects (A.I.A.) is their primary trade association. Find one that specializes in residential design rather than a shopping mall designer, for obvious reasons. I'll show you how to find and draft a great architect (as well as material suppliers) in Chapter 15.

On the Level

Don't know which architect to hire? Ask local builders and the building permit department who they use. Will they recommend someone? Why?

As mentioned in Chapter 4, architects can be paid by the time (typically $50 to $125 an hour) or as a percentage of the total project value (8 to 15 percent). For their fee they can develop the design, draw the plans, produce construction documents, and even help manage the construction. You get what you pay for.

If you do opt for architectural services, know that everything is negotiable, especially if you're willing to participate. For example, bringing in a comprehensive Home Book on your first visit will save time and money. Ask how you can reduce costs while still maintaining quality.

One way of lowering costs can be to bring in stock plans (you'll learn more about them in a moment) for additional work. This can save an architect time and you money.

Another is to ask the architect if they offer stock plans of their own that can be modified to fit your design. This, too, lowers your costs.

Even if you buy stock plans, consider having a local architect look them over to make sure they can be built under local code and construction practices. A few hundred bucks spent here may save thousands later.

Of course, if you're going to buy a kit house, all the needed plans will come with it—or should.

House #12940267: Using Stock Plan Services

Earlier in this chapter you were caught looking through stock plan magazines, catalogs, and books. Maybe you saw one that fit your design requirements. Or maybe it was pretty close.

Stock plan services can dramatically cut the cost of designing your own home. For example, a full set of plans often can be purchased for $500 to $1,000. For just a couple hundred more you can have numerous copies to give to the building department and all contractors involved.

What's included in a typical stock plan package?

- ◆ Construction plans
- ◆ Plumbing plans
- ◆ Electrical plans
- ◆ Mechanical (HVAC) plans
- ◆ Materials list
- ◆ Specification outline
- ◆ Summary cost report
- ◆ Materials cost report
- ◆ Additional house planning aids

Ka-ching!

You can save thousands of dollars by selecting a stock plan and having it modified by an architect or drafting service to fit your needs.

Most of the stock plan services can even localize the plans and the cost reports based on the Zip Code in which you're building. Package enhancements can be purchased that include suggested landscaping and decking. Also, most stock building plans can be reversed. That is, the right side of the plan can be on the left, or the top can be on the bottom. Ask for right-reading so you don't have to hold the plans up to a mirror to decipher them.

An architect may charge 2 to 3 percent of the project's value for a full set of plans. That's $5,000 to $7,500 for a $250,000 house. Stock plans offer a cost-cutting option. Appendix B includes contact information on some of the major stock plan services.

Volunteering for a Drafting Service

You can draw and build from your own self-drawn plans. The problem comes when someone needs to interpret your plans. ("What was he thinking?!")

If you're going the do-it-yourself design route, consider hiring a drafting service or architectural illustrator to develop the plans. (In most areas, you *must* have plans drawn to local building standards to get a building permit even if you build the entire house yourself.)

Most drafting services can work with your sketches, adding the detail needed by tradespeople and the permit department. Most will be drawn using computer-assisted design (CAD) software, which means that changes are easier than on hand-drawn plans. Changes will still cost, but not as much.

You can find drafting services in the business section of your local telephone book.

Using Design and Plan Software

The computer has revolutionized many industries, not the least of which is home construction and remodeling. Professionals have powerful tools for sketching, sizing, placing, and designing everything from small rooms to housing complexes. Fortunately for consumers, many of these powerful (and expensive) tools have consumer-level counterparts that are inexpensive (under $100) and quite useful in designing and planning projects like building your own home.

Ka-ching!

Before ordering stock plans from a service, ask for a free or low-cost sample package to help you determine the quality of their designs and plans. Take them to a builder or architect for comments, remembering their own bias. The *right* stock plans can save you thousands of dollars. The *wrong* ones can waste thousands in building costs.

Professional Home Design (PHD) distributed by Punch Software (www.punchsoftware.com), for example, is a popular home construction and remodeling design program that is actually a family of programs. PHD allows you to start a design by quickly drawing the perimeter of the foundation, then adding the flooring system, electrical, plumbing, roofing, HVAC (heating, ventilation, and air conditioning), and even landscaping. Once you've placed all the walls, doors, and other components, it can produce electrical, plumbing, and HVAC plans for your printer in seconds. PHD and many other consumer design programs also can develop and print materials lists for projects. You can use these lists to get bids from suppliers.

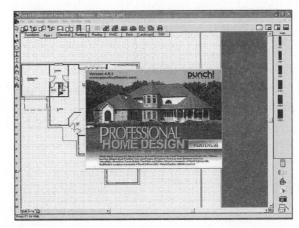

Computer screen for Punch Professional Home Design program.

However, one of the most popular components of this and other home design and remodeling software packages is the three-dimensional or 3D view. Once your plan is completed, you can virtually walk through your new house on the computer screen, turning left or right as needed. In addition, many of these programs can instantly strip away the surfaces to expose floor joists, wall studs, and ceiling rafters.

Other consumer home design software includes myHouse from DesignSoft (www.designsoftware.com), Classic Home Design from Artifice (www.artifice.com), SmartDraw (www.smartdraw.com), and 3D Home Architect developed by Advanced Relational Technology and distributed by Broderbund Software (www.broderbund.com).

Ka-ching!

Make sure the remodeling and design software you select is appropriate to your needs. Don't purchase one that will take weeks to learn and even more time to use. If you have Internet access, download trial copies of various design programs to determine which is best for your needs.

On the higher end are products like Chief Architect, also from Advanced Relational Technology (ART) and sold direct (www.chiefarchitect.com). Developed for architects, builders, designers, and drafters, it is more expensive than consumer design products—but you get what you pay for. The library that comes with it includes 7,000 symbols, textures, and images to make a walk-through almost feel like you've broken into someone's home. You can even specify a door or window by manufacturer and size, for example, and plop it into your design to see what it will look like. Once you're done, you can have the program print out a one-dimensional model that can be cut and assembled into a three-dimensional room or house. Talk about seeing what you're getting into!

Computer screen for ART Chief Architect professional software.

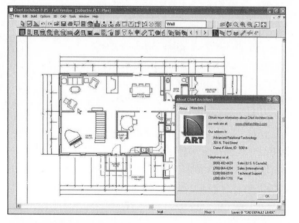

Three-dimensional house plan.

There are numerous other features available in consumer- and professional-level home remodeling design software. There are many other software programs available as well. Larger computer software stores will have a few on hand. Otherwise, check online, download trial versions, and have some fun designing your new home.

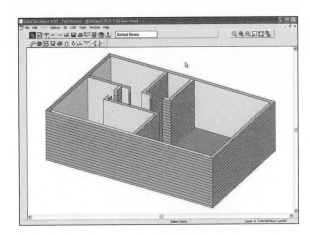

Three-dimensional foundation plan.

Not everyone has—or even wants—a computer. If you'd prefer to draw out your own plans, stop by a stationery store for graph paper, mechanical pencils, and rulers. You can use each square of the graph paper to represent 1 foot of your house, meaning a 20 foot by 30 foot house will be 5 inches by 6 inches on paper.

Analyze This!

Here's the question: Can you efficiently build from the set of plans you bought or drew? Your local building permit department can look at the plans and tell you if it conforms to local codes.

You can answer this question yourself if you're willing to do some homework. There are numerous books on the structural principles of residential construction. Some are included in Appendix B of this book. In addition, local building code manuals will include structural guidelines.

Most folks doing their own design and plans consider hiring a structural engineer to look the plans over before building.

If you're still interested in saving some money, stick around for the next chapter on building an energy-efficient home.

Code Red

Numerous changes to a plan set can make for a messy—and inaccurate—plan. In one case, last-minute roof changes weren't carried over to the floor plan, making expensive and preventable errors.

The Least You Need to Know

◆ Gather lots of design ideas before you start deciding what's best for you.

◆ Figure out how much you have to spend before you decide what your home will include.

◆ Be prepared to make compromises.

◆ You can save hundreds or thousands of dollars by developing your own house plans.

◆ Hire the best architect or plan service that matches your design.

Chapter 6

Making Your Home Energy Efficient

In This Chapter

- ◆ What makes a house energy efficient?
- ◆ Understanding heating and cooling systems
- ◆ Ways to cut long-term energy bills
- ◆ Finding energy-efficient materials for your home

Building your home isn't cheap. And maintaining a comfortable climate in it can be expensive. Fortunately, building your home "energy smart" can reduce the ongoing costs and save thousands of dollars over the life of your home.

So that's the topic we're going to tackle in this chapter: home energy efficiency. You can do so much right now, as you plan and design your home, to take advantage of the latest technology and ideas. Even without extra equipment, smart designing can dramatically cut your utility bills for years to come.

So pull out your handy-dandy Home Book and learn how to make your home warm and cozy, or cool and comfortable, at the lowest cost.

Back to the Drawing Board: Designing for Efficiency

Here's how it works: Heat energy seeks equilibrium. That is, heat on the outside of a cooler house wants to get in, while heat on the inside wants to go outdoors and play. If just thick walls stood in its way, heat transfer would be very slow. But most houses have doors and windows and lots of little air leaks to speed up the transfer.

Of course, that wouldn't be a problem if climate maintenance were free. Unfortunately, keeping a house cool in the heat and warm in the cold costs lots of bucks. Once purchased, you want that high-priced warm (or cool) air to stay where you put it. The trick is to minimize heat transfer through the house's walls, doors, and windows. (Or buy a *really* big heating-and-cooling system and purchase stock in the local power company.)

Fortunately, you can, at relatively little expense, design your house to minimize heat transfer. People know lots more about how to do this than they did when your parents shopped for housing. What you're looking for as you consider which building materials to buy are those materials with the highest *R value*.

> **Building Your Vocab**
>
> **R value** is a laboratory standard that defines a material's resistance to heat transfer. The higher the number, the more it will resist the transfer of heat through it.

Of course, if some material has a specific R value, more is better. That is, 4 inches of fiberglass has an R-11 rating. Six inches is rated at R-19. That's why 2" × 6" exterior walls are more energy efficient than 2" × 4" walls.

Local code and building practices will dictate the minimum R values you should have in your house. Obviously, you don't need the same R value for a home in San Diego that you do for one in Minneapolis. There's a point of diminishing returns, however. For example, it may cost more to increase the R value to its upper limits than the amount of energy it will save. Smart planning is important.

Here are some ideas to consider as you plan your house to be more energy efficient:

◆ Ask your primary energy supplier about an energy audit of your house design. Many have on-staff experts who can review your rough plans and make recommendations on energy efficiency.

◆ If you need a higher R value, build your house with 2" × 6" exterior walls and pack it with thicker insulation.

◆ Use the most efficient insulation you can afford.

◆ Install low-emissivity (low-E) glass coated to reduce heat conductivity without reducing light.

◆ Insulate ceilings well, even higher than the local recommended R value.

◆ Well-insulated ceilings mean that you should ventilate the attic, to allow trapped heat to escape.

◆ Check walls as they are built, and fill up any air leaks from the outside.

◆ Include a moisture barrier in walls to keep them from sweating.

◆ Heat rises, so install heat registers near the floor and the return vent near the ceiling.

◆ Fireplaces are very inefficient heat sources (see the upcoming section). If you must have one in your home, buy the most efficient model you can afford.

◆ Buy a heating furnace rated in *BTU*s and energy efficiency to match or exceed your house's needs.

> **Building Your Vocab**
>
> A **British Thermal Unit** or **BTU** is a measurement of the quantity of heat. One BTU is the amount of heat needed to raise the temperature of one pound of water one degree Fahrenheit, from 59°F to 60°F. Heat loss is measured in BTUH (BTUs lost per hour) units.

◆ Include ceiling fans and other circulation equipment in your plans, to efficiently use heated and cooled air.

◆ Make sure you install a computerized thermostat that you can set to make multiple temperature changes during the day. Every degree you turn down your heat setting or up your air conditioning setting can save you about 2 percent of your fuel bill.

◆ Include double-glazed windows in your budget for long-term energy savings.

◆ Insulate any heating ducts that run through unheated basements or crawl spaces.

◆ Remember to use weather-stripping and apply caulking around doors and windows.

Understanding Heating and Cooling Systems

One of the most important choices you can make toward an energy-efficient house is selecting the most efficient heating and/or cooling system for your location.

Chapter 22 includes instructions on installing common heating, ventilating, air-conditioning (HVAC) systems. The decision of which one comes in the design process because it will impact how you build. Let's take a closer look at your options.

Turn Up the Heat: Heating Systems

One of the most popular residential heating systems is called direct heat. Air is heated in a furnace and circulated through the house using large pipes called ducts. The air exits into the rooms through registers that you can regulate. Cooler air returns to the furnace through a cold-air return.

Direct heat systems are typically fueled by oil, propane, natural gas, or electricity. Which one is the most cost efficient depends on how close you live to the source of this energy. Transportation costs money. Transporting heat in your house, through ducts, also costs money, so plan for the heating system to be as centrally located as possible to minimize transit loss.

Convective heat uses baseboard heaters. Most modern systems circulate hot water that heats the surrounding air convectively. Others use electricity. Radiant heat warms the interior atmosphere by heating the floor or ceiling. Hot-water pipes or electrical wires are installed in the floor or ceiling during construction.

A heat pump is a combination heating and cooling system in one unit. It releases heat to warm the air, or it absorbs heat to cool the air. Pretty simple. Heat pumps aren't as practical for extremely cold climates, so auxiliary heating systems are added.

Baby, It's Cold Inside: Cooling Systems

Traditional cooling systems use refrigerant to chill the air, which is then circulated. Central air systems distribute the cooled air through ducts, probably the same ones that are used for the heating system. Room air conditioners directly circulate the cooled air into the room.

HVAC for Owner-Builders

Unfortunately, heating, ventilating, air-conditioning (HVAC) systems are a specialized trade. These systems are typically not manufactured for owner-builders. However, if you want to do it yourself, start hitting the books now to learn how they are installed. And make sure that whoever draws your house plans includes HVAC details or even a separate plan. (Because of the variety of heating systems nationwide, many plan services only include rudimentary HVAC plans.) Alternatively, plan on hiring an HVAC expert for the installation. You can, if you wish, help with the framing and installing of duct work and save some dough.

Old-Fashioned Heat: Fireplaces

Though wood-burning fireplaces, too, have become much more efficient over the past 20 years, they may not be allowed in new home construction in your area. Better check with the local planning department first. Home fireplaces can be very polluting as well as an inefficient source of heat.

Your fireplace should have some built-in method of efficiently circulating the warmed air. How it all works depends on the model you select.

If you must have a fireplace and it is only for ambiance, consider a gas fireplace with ceramic logs. They are more efficient than wood-burning fireplaces if they include an air circulation system.

 Ka-ching!

If you can have one, consider a zero-clearance fireplace. It's a prefabricated metal fireplace that you can encase in a frame and masonry to look like a traditional fireplace. It's much more efficient than an open fireplace.

Natural Heating and Cooling

Technology is quite a tool. We've learned much about how to take advantage of the sun toward making homes more comfortable year-round with the least amount of artificial heating. The technology is called passive solar. Passive solar heating has three major applications:

- **Indirect gain** uses large blocks of something to collect and store the heat throughout the day, releasing it in the cooler night. The "something" can be adobe, concrete, glass, water, sand, or any other objects with slow thermal loss. The mass is built into south-facing walls to collect daytime heat for later use.
- **Direct gain** gathers solar energy for more immediate use. Most systems use south-facing glass walls to allow the sun's rays to enter.
- **Isolated gain** systems collect and store heat in an area thermally isolated from the living area. As needed, the energy circulates to the rest of the house.

In addition, there are some really smart ways of cooling your home passively. Natural ventilation takes advantage of air flow on your building site, adding windows and other barriers to assist the flow.

Evaporative cooling makes living spaces more bearable by adding water to the air. In some humid locations, however, that's the last thing you want. You'd rather take water from the air to decrease the humidity. In that case you want desiccant cooling.

Let's Get Active: Solar Heating and Cooling

Passive systems don't rely much on fans or blowers to circulate the treated air. If they do, they're considered active systems. The most popular solar heating system uses solar collectors, large panels of pipes that intercept solar radiation and convert it into heat for storage and use. For example, a solar water heating system circulates water in roof panels, and then uses it for heating air or for hot showers.

Thermal storage units are large chambers of air, liquid, or rock that gather and store heat for mechanical distribution using fans. Indirect systems gather energy in collectors mounted on the roof, and then process the energy for use in various ways within the home. As with passive systems, you can use active solar heating and cooling systems as the primary source of climate control or as a secondary or auxiliary source.

This is just the tip of the melting iceberg. There are thick books on how to include active and passive heating and cooling systems in the design of your house. Some are included in Appendix B, and I'll give you additional resources at the end of this chapter. In addition, refer to my book *The Complete Idiot's Guide to Solar Power for Your Home* (see Appendix B) for more comprehensive help.

Choosing the Best HVAC System

Which is the best HVAC system for your home? Obviously, you have lots of options: fireplace, electrical, gas, solar, heat pump, etc. However, your choices may be limited by what's available and what's most efficient.

Some home sites, especially those in rural areas, may have limited sources of fuel. If your new home is "off-grid" or not connected to a municipal or commercial power source, you may be limited to solar, wind, or other alternative energy. In other locations, electricity is less expensive than other resources such as natural gas or even propane.

The best place to start when choosing an HVAC system is with an architect or heating consultant. They can tell you what is available, the relative costs of installation, operation, and maintenance, and other factors that you might not learn without professional help.

If you are considering a wood-burning fireplace, also consider what your source of fuel will be. Will you purchase wood pellets and have them delivered? Are there reputable sources for seasoned firewood? If so, can you physically handle the bags and logs? Should you have a backup source of fuel such as a gas furnace or electric wall heaters?

Many public utilities offer an energy audit for planned homes as well as those already built. They can help you estimate the heating needs and your resources. They also can help you estimate whether a specific source is expected to dramatically increase in costs over the next 10 years.

Saving Energy with Smart Materials

Insulation simply slows down the heat transfer process. Insulation is good. There are many popular types of insulation, including the following:

◆ Flexible insulation

◆ Loose-fill insulation

◆ Reflective insulation

◆ Rigid insulation

◆ Everything else

Your choice depends on local building code, availability, and which can be most efficiently installed in your home.

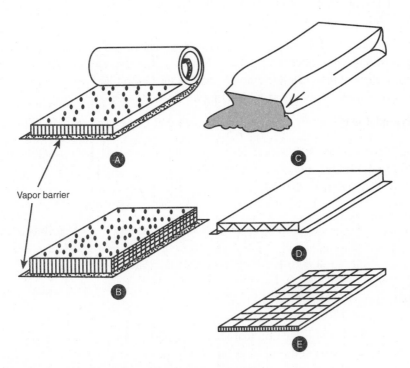

Popular types of insulation include (A) blanket, (B) batt, (C) fill, (D) reflective, and (E) rigid.

Vapor barrier

Blankets and Batts, Oh My!: Flexible Insulation

You've seen rolls of flexible insulation before. Typically, it's made of fiberglass or of similar fibers. It comes in two types: blanket and batt. Not much difference between them.

Blanket insulation comes in rolls or packages in widths suited to 16- and 24-inch stud-and-joist spacing. One of the more popular brands is pink in color. The body of the blanket is made of felted mats of mineral or vegetable fibers such as rock or glass wool, wood fiber, and cotton. Organic insulations are treated to make them resistant to fire, decay, insects, and vermin.

Most blanket insulation is covered with paper or other sheet material and has tabs on the sides for fastening to studs or joists. One covering sheet serves as a vapor barrier to resist water vapor. It should always face the warm side of the wall. Aluminum foil, asphalt, or plastic-laminated paper are commonly used as exterior-barrier materials.

Batt insulation is made of fibrous material in thicknesses of 4 and 6 inches for 16- and 24-inch joist spacing. It is supplied with or without a vapor barrier (see the previous illustration). Friction batts are supplied without a covering and will remain in place without staples.

Getting Loose: Fill Insulation

Loose-fill insulation comes in bags or bales, and you place it by pouring, blowing, or packing by hand. It includes rock or glass wool, wood fibers, shredded redwood bark, cork, wood pulp products, vermiculite, sawdust, and shavings.

Loose-fill insulation is normally used between first-floor ceiling joists in unheated attics and around flat-roof skylights. It is sometimes blown into walls and around doors and windows.

Bouncing the Heat: Reflective Insulation

Most materials reflect some radiant heat, and some materials do so more than others. The more reflective materials include aluminum foil, sheet metal with tin coating, and paper products coated with a reflective oxide.

Reflective insulation is used in enclosed stud spaces, attics, and similar places, to retard the transfer of radiated heat. It is effective only when used where the reflective surface faces an airspace at least ¾ inch or more deep. Foil-type reflective insulation is often applied to insulation blankets.

By-the-Sheet: Rigid Insulation

Rigid insulation is usually a fiberboard material manufactured in sheets. Structural insulating boards range in densities from 15 to 31 pounds per cubic foot and are fabricated into building boards, roof decking, sheathing, and wallboard. While they have moderately good insulating properties, their primary purpose is structural.

The most common form of rigid insulation is sheathing board in ½-inch and ²⁵⁄₃₂-inch thicknesses. It is coated or impregnated with asphalt or some other compound to provide water resistance. Sheets are 2' × 8' for horizontal installation, and are 4' × 8' or longer for vertical applications.

Other Insulating Stuff

There are other energy-efficient home insulation materials available, with new ones coming out all the time. They include insulation blankets made of multiple layers of corrugated paper.

Other materials are formed-in-place insulations, which include sprayed and plastic types. Sprayed insulation is usually inorganic fibrous material blown against a clean surface that has been primed with an adhesive coating.

You can mold or spray foams in place; you can apply urethane insulation by spraying. Polystyrene and urethane boards are available in ½-inch to 2-inch thicknesses.

On the Level

Place insulation on all outside walls and the ceiling. In houses with unheated crawl spaces, install it between the floor joists or around the wall perimeter. Support flexible blanket or batt insulation between joists by slats and a galvanized wire mesh or by a rigid board with the vapor barrier installed toward the subflooring. Press-fit or friction insulations fit tightly between joists and require only a small amount of support to hold them in place. Reflective insulation is often used for crawl spaces. You can place a ground cover of roll roofing or plastic film such as polyethylene on the soil of the crawl space to decrease moisture content.

Choosing the Best Insulation

Which type of insulation should you use in your home? The various types of insulation are manufactured to meet specific needs. For example, loose fill insulation often is installed in spaces that are already enclosed with drywall. Flexible insulation (batt and blanket) are typically the most popular with new home construction. Rigid insulation can be installed over thinner walls or to add an extra layer of insulation in colder climates.

Your architect or your local utility provider can help you select the insulation that is most efficient for your new home. In addition, your materials supplier can suggest which are the easiest for homeowners to install. The most popular insulation is blanket for cost, efficiency, and ease of installation.

Using the Latest Ideas

The United States has numerous climates and microclimates within it. Here's how you can learn what energy-efficient systems can work best for your house:

- ◆ Know the comparative cost of energy in your building location. Will it change in the future?
- ◆ Contact your energy supplier (gas, oil, electricity) and ask for local resources for new construction.
- ◆ Visit local and regional home shows for the latest ideas and literature, asking about installation steps.
- ◆ Review Appendix B for additional resources on energy-efficient building.

Selecting the most efficient building area and site is an important part of the building equation. Chapter 7 covers selecting a location, and Chapter 8 is about picking the best site to build your own home.

The Least You Need to Know

- ◆ Your home's design is the best place to start saving money on energy for years to come.
- ◆ A material's R value notes its resistance to the transfer of heat; higher numbers are better.
- ◆ Many energy-efficient materials are available for making your home more comfortable for less money.
- ◆ Contact local energy suppliers for help in planning your new energy-efficient home.

Chapter 7

Where to Live: Choosing a Location

In This Chapter

◆ What are you looking for in a location?

◆ Considering location options: urban, suburban, or rural

◆ Learning about the best locations

◆ Understanding neighborhood restrictions

◆ Making location compromises

So where are you going to plant your new house? It's an important question! In fact, this whole chapter is dedicated to answering that one question. Not the specific lot; that's covered in Chapter 8. You first need to identify a group of buildable sites from which to choose.

Why is location so important? Because where you build your home has extensive impact on how much you enjoy it and even how much it's worth. Location can impact your lives as much as the house itself.

You'll see what I mean as we walk through the process of selecting your new home's location.

Where Do You Want to Live?

Before you start thinking about your location options, take some time to consider what you want from the location of your home. Here are some thought-provoking questions:

◆ How will you use your home over the next 10 years? Starting and raising a young family? Helping teens move on? Preparing for retirement?

◆ How near to public transportation (buses, trains, airports) do you need to be?

◆ How near to health or professional services do you need to live?

◆ How near to recreational activities (camping, golf, club) do you want to be?

◆ How much time do you have to select the best location for your house?

◆ What services do you need delivered to the site (sewer, water, power, gas) and which do you prefer to develop on-site (septic, water, solar power)?

◆ Where are better schools located and where must you live to have your children enrolled?

◆ How large of a lot do you need to live comfortably?

◆ How much of a factor do crime rates play in your decision? (No one wants to move into a crime-ridden neighborhood, but sometimes you can make a difference—or at least install an alarm system—to get into the area you want.)

◆ What view would you like out of primary windows?

◆ How do you picture the ideal neighborhood?

◆ What ideas do others in your household have about where to live?

Start making notes in your Home Book.

On the Level

Need to find out more about somewhere you don't live? Check local libraries for a copy of *Moving & Relocation Sourcebook and Directory* (Omnigraphics, updated annually). It includes extensive contact listings for more than 120 U.S. cities. At over 1,200 pages and $200+, it may not be in all libraries. Get the latest edition if you can. It's a great resource.

Ka-ching!

Want to build in the city but no lots are available? Consider buying a good lot with an older structure that can be torn down. You may even be able to use the existing foundation.

Considering Your Location Options

What are your options? The most obvious are urban, suburban, or rural. Grab your Home Book and your significant-other-decision-maker to start reviewing and discussing your options.

And remember to keep an open mind. Even if you've already bought your building lot, consider all of your options before committing to building on it. You may find the home of your dreams was moved.

Summer in the City

Urban means in the city. Of course, the "city" may be more like a large town, or it may be a metropolis. So consider what works best in your life. Is living near an urban job, arts and entertainment, and nightspots important to you and your family unit?

A primary decision element is the commute. Not just the trip to work for those who hold down jobs, but also the necessary trips your family makes. That means proximity to schools, worship centers, popular activities, shopping, and other stuff. Figuring what your commute(s) will be means calculating time *and* mileage.

You may discover that although an urban location isn't ideal, it's the best option for your next house. It fits your employment and family needs the best.

Many families have moved away from urban areas because of higher crime. And it is still true in many cities. However, changes in law enforcement and urban planning have reduced crime in many others. If an urban residence otherwise makes sense, don't rule it out without learning more.

Moving to the 'Burbs

The 1950s began urban sprawl as folks working in the city decided that they could easily commute from the suburbs or "burbs." Unfortunately, *everyone* seemed to have the same idea, and roads became clogged with commuters each morning and afternoon. Then it was 24 hours a day.

You can ask the same question: What location best fits your lifestyle? Here are some other important questions:

- Do you have children who will benefit from a specific suburban school or activity?
- Is your urban employer planning a move to the suburbs or the edge of the city?
- Do you have more employment opportunities in another area that's near a specific suburb?
- Do you find certain suburbs in your area to be more desirable than others?

Suburban lots will probably be more readily available and lower in cost than those in the city. Utilities such as water, sewer, and electricity will be delivered to the lot, unlike some rural building sites.

Suburbs are residential communities that serve commercial communities. However, most suburbs have evolved to include their own commercial and service centers called shopping malls. The latest trend is toward *boomburbs* that include as many as 100,000 residences but primarily rely on commercial centers outside their borders. They are truly "bedroom communities."

Goin' to the Country

A home in the country! It's many people's dream, and it can be yours. Or it can be a nightmare. The difference is learning as much as you can about rural building. For example, services are delivered to suburban lots, but in the country, you may need to build the services onsite or bring them in from afar. The price of a rural lot may be lower than one in the suburbs, but it may need a septic system, a water system, and maybe even an electrical generation system to offer the kind of life you are accustomed to. Take a look at my book *The Complete Idiot's Guide to Solar Power for Your Home* (see Appendix B) for ways to use alternative energy for better living.

You can build rural houses without electrical service. In fact, in some areas it's not even an option. General contractor Greg Dunbar, this book's technical advisor, lives in a rural home that's "off grid." I'm not as brave. Our home is in a rural subdivision with sewer, water, and electrical service to the lot line.

Again, consider your current and future lifestyle before committing to a rural house. It may be just what you want. Or it may be a long dusty-road commute to the grocery store.

What Are You Moving Into?

Once you've decided whether your next address will be urban, suburban, or rural, you need to start looking deeper. You'll have many questions to ask:

- Where are the better building areas?
- How much do typical building sites cost?
- Is the price all-inclusive or will it need additional services?
- How are the local schools rated academically?
- Whassup in the 'hood? (Translation: What changes are occurring in the neighborhood?)
- Who are the major employers in the area and how stable are they?
- What's the local property tax structure and how might it change?
- How well managed is local government?
- How stringent are local building codes?
- Are local covenants (discussed later in this chapter) enforced?
- If wage earners lost their jobs here, are other jobs available for them nearby?
- Is this a place that you and your household would enjoy and thrive in?

Numerous resources can answer your questions about a specific area. Real estate agents and brokers are your first contact. Experienced professionals can answer many of your initial questions and probably many more that you'll think of. So start calling around to agents in the areas that you're considering.

Want to find out which real estate agent can best help you? Contact a dozen or more with the exact same request, and watch who follows up and how. Something like: "I'm looking for a buildable lot in the Brookside neighborhood." Better agents will spend some time with you finding out more about your specific needs and then following up. Others will not. Choose accordingly.

Code Red

Remember: Lower property taxes typically mean less education and protection services.

Another good source of information about a specific area is people you know who live there. Have a party for all your friends and acquaintances, asking, "Does anyone here live or know someone in Brookside?" for example. (Good excuse for a party!) Try the same question at work and at school.

During the building process, you're probably going to have to introduce yourself to a lender. Why not start now? Speak with local lenders, asking questions about your potential neighborhoods. One may even live there. And you may make a financial friend.

On the Level

If an employer is relocating you and you're planning to build a house at your new location, find out how your employer's human resources department or relocation service can help. Also, find a copy of my book *The Complete Idiot's Guide to Smart Moving* (Alpha Books, 1998).

Local newspapers can help you learn more about the issues in your new location. Subscribe to and read neighborhood or nearby city newspapers. Or visit the newspaper office and ask to browse their library for stories about your chosen area. They may be indexed and on microfiche, making your job easier.

The chamber of commerce for your destination can also be helpful. Some have valuable packages of information they offer to those moving into the area. Others will give you a generic brochure, but answer specific questions like "Where are area elementary schools located?"

If you're plugged in to the Internet (or know someone who is), search online about the area(s). Many communities have their own websites including the minutes of public meetings and even water quality reports online.

Police departments can also tell you about crime rates or crime issues in specific areas. They may not want to do so over the telephone, but you can often find a community-relations officer who will talk about such things in person. Police department websites may post local crime statistics, or at least provide a contact name to whom you can e-mail specific questions.

I'm Your New Neighbor!

Riffraff, keep out! That's the sign that some neighborhoods would love to post at their borders. And you can't blame them. Folks spend lots of time and money buying a house in a neighborhood they enjoy. They want their neighbors to enhance that investment.

If you plan on building in a community that you're not familiar with, invest in a trip there—and preferably several—to scope things out. Call ahead and find a real estate agent who will show you around. Meet with school, church, lodge, and other folks you'll be getting to know later. Better to do some legwork now than be faced with any unpleasant surprises later.

Zoning Issues

Actually, zoning starts at the county or city level. Cities enact zoning laws to isolate land uses. That is, they want to keep commercial uses separate from residential uses. Imagine building your new home and finding out afterward that a Wal-Mart was going in next door. (Hey, maybe that's not such a bad idea … but think about it!)

Zoning laws across this great nation could take up a whole book unto themselves—a really boring book. So I'll summarize here. Agricultural zoning typically means one major residence (and maybe some worker housing) per tract of land. And you'd better be growing something legal. In some cases, a personal garden qualifies as agriculture, while in other locations it means that a combine should be parked in the barn.

Don't expect to buy up a large parcel of land zoned for agriculture and chop it up for residences. That day is just about over. It takes land-use experts to get large parcels rezoned for residential use. Maybe you're living in an area where there are exceptions, so it's worth a call to the local planning office to find out. Just don't be surprised when they chuckle.

Environmental zoning means no one has ever built on it before (except maybe Native Americans) and probably no one ever will. So forget it.

Commercial and industrial zoning means that you can build a business or a manufacturing plant on the land. In some areas, you can also build lower-use structures in commercial and industrial zones. That means a house. But why? The only two reasons would be these:

- You didn't really care who your neighbors were.
- You were planning to use your residence for commercial or industrial use as well.

Which brings us to the most popular residential zoning group, called—ta-dah!—residential zoning. Actually, most areas have more than one residential zone category. What they have in common is how you use the property. It's for structures that people live in. Makes sense.

The subcategory zones tell how many residences can be on the property, or how many acres they require per residence, or some other criteria. So, an R-2 zone may mean two residences per site, or a two-acre minimum per residence, or a residence that can only be occupied by two-year-olds—whatever the local planning boards want it to mean.

The point here is to find out what the zoning rules are in your neck of the woods. Discuss them with a local real estate agent, the county or city building department, or someone in the planning department. Know what you can and cannot do before you try to do it.

Covenants Are Our Friends

A covenant is an agreement. Neighborhoods, developments, and subdivisions often have covenants (sometimes called "covenants and restrictions," or "C&Rs") to keep out the perceived riffraff.

It's usually the land developer who writes the covenants, depending on whom it is selling the land to. If it's for high-end homes, the list of covenants and restrictions will be longer, such as the following:

- A review board must approve home plans.
- Homes must be of a designated size or larger.
- No on-street parking is allowed.
- Roofs must be of a designated type and quality.

 Ka-ching!

Save time and money by finding out about C&Rs early. And don't expect you can change them easily. Accept them or move somewhere else.

◆ Existing trees cannot be removed without approval.

◆ Vehicles parked overnight must be in enclosed shelters.

◆ Trash containers must be kept in enclosures.

◆ No farm animals or large exotic pets are allowed.

So who are the covenant police? Everybody! Anyone who owns property within the same subdivision can make a complaint about an infraction to whoever's in charge. Darn! That means it's really important for you to read and understand the C&Rs for the area where you plan to build. They will be minimal or even non-existent in agricultural zones, and will be four miles long in exclusive high-end neighborhoods.

That's okay. If you put up a $500,000 house, you don't *want* the Joads parking their dilapidated truck out front and living on the lot next door. And you may not want to wake up to roosters in the morning. (If you do, consider an agriculture zone.)

Who actually enforces C&Rs? Because they are private contracts between those who buy land in a designated area, C&Rs don't fall under the jurisdiction of the courts—unless they're illegal contracts. A property owners' association (POA) typically enforces C&Rs. The power and structure of the POA is outlined in the C&Rs or in related documents typically available to the public.

On the Level

Covenants that restrict people from living in a neighborhood because of their race, creed, religion, or national origin are unconstitutional, and thus illegal. As they should be.

How can you find out if property you are considering is covered by covenants and restrictions, and what they are? First, try the POA. If possible, talk with someone from the group to learn how the subdivision came about and what its goals are. Then get a copy of the C&Rs and read it thoroughly.

You can also get a copy through most local title companies. It's their job to keep track of ownership, sales and transfers, and property rights, as well as covenants and restrictions. There may be a small fee. Check the local phone book for title companies, or ask your real estate agent to get the info for you.

Covenants and restrictions may expire. Some have a specific life of 20 or 30 years, while others expire and should be replaced once all lots in the subdivision are sold. Also, there may be C&Rs on specific lots in the subdivision but not on others. Find out.

Architectural Review Board

As I mentioned earlier, some planning organizations and covenants require that homes built within its jurisdiction first pass an architectural review board (ARB). The reason is that the covenants, and all property owners when they bought in, agreed that houses should meet specific standards. It may be that they all must be of a similar architectural design. Or the rooflines must be complementary. Or exterior colors must be pastels. Or the size must meet a minimum square-footage.

So it's a good idea to get copies of the review guidelines before drawing up plans. If review is required, it will probably be by a local designer or architect who works with the property owners' association or some other board. Talk with that person in advance of finalizing your plans. You may decide to make some necessary changes. Or you may decide to build elsewhere.

If you'll be working with an architect or residential planner, she or he can help you make sure that your design meets the ARB's requirements.

I'm Okay, You're Okay: Making Compromises

Life is a compromise. So is just about every major decision we have to make. Building your own home will include numerous compromises (like "champagne on a beer budget").

However, the location of your house should not be an easy compromise. Hold out for the best you can afford, cutting corners in other places as needed so you can have the location that best fits your lifestyle. After all, you can always upgrade your home later, but you can't easily move your home to a better location.

The Least You Need to Know

◆ The location you choose for your home is one of the most important housing decisions you will make.

◆ Choose between urban, suburban, and rural locations based on what works best for you and your household over the next few years.

◆ Learn about neighborhood covenants and restrictions *before* you select your building lot.

◆ Be ready to make a few compromises to get as much as you can of what you want in a location.

Chapter 8

Selecting a Building Site

In This Chapter

♦ Finding available building sites

♦ Selecting the best site for your house: questions to ask before you buy

♦ Estimating site costs

♦ Buying the best site for the best price

You've narrowed it down. With careful thought and research, you've decided on one or two specific areas where you may want to build your own home. You've considered needs, lifestyles, zoning, restrictions, and even made a compromise or two. Good going!

The location of your home is *so* important that this book devotes *two* chapters to the topic. The previous chapter, Chapter 7, helped you find the best area and type of lot for your home. This chapter guides you to the best available building site and helps you figure out how much it will *really* cost to build on it.

With your Home Book in hand, let's start the hunt.

Where, Oh Where?

Okay, you know you want to consider building sites in Brookside or in Oak Manor, for example. Now what? You've probably already driven the streets of these two areas until you know them by heart. If not, you should. One of these very well may be your next neighborhood. As you drive around looking at available lots, also look for these things:

♦ Real estate sales signs

♦ For Sale By Owner signs

♦ Built By signs

♦ Modified lots

♦ Preferred and unpreferred locations

♦ House designs

Maybe you've already asked a real estate agent to help you find your building lot. If not, considering doing so. Contact the agents listed on For Sale signs. If you already have an agent, give her or him information about the lot, to follow up.

For Sale signs can tell a story. Carefully look over signs to determine how long the property may have been available. Faded FSBO signs are clear indications of long-term availability. Since agents reuse signs, they may not be as indicative, so study the ground where the sign is planted to guesstimate how long it's been there.

You'll also see For Sale By Owner, referred to as FSBO, or "fiz-bo," signs. Jot down the information, and contact the owner.

Builders may have signs out in your selected areas as well. They can be a good source of local builders who have experience in the neighborhood. If under construction, you can actually see how well or poorly they are building the house, and can even talk with tradespeople. You may just learn who *not* to hire.

In addition, builders often buy numerous lots in a subdivision so that they are assured of future work. However, they may decide (for financial or other reasons) to not build on all of them. Some lots may be for sale even though there's no For Sale sign. Builders may even sell prime lots because they need the cash for another project.

Modified lots can be a source of building sites. Here's what I mean. A lot, instead of being cleared for construction, may have some excavation that has since grown over. That probably means someone considered building on the site, even started, and then stopped. Of course, it may also mean there's a problem with the lot. Or it could mean that plans changed. The lot may again be available for sale.

Ka-ching!

You may be able to get a free (or low-cost) package of public-record information on any property in your city or county. Where? At a local land title company. Call and ask. Or try your city or county public records department.

How do you find out who owns it?

Go to the local courthouse. The records or tax assessor's department should have the information you need. They can give you the name and tax address of the owner of any property in the county. It's easier if you have a subdivision name and lot number, but most can get it from an address (or surrounding addresses). It's public record; however, they may charge you a fee for the labor of looking it up and printing copies for you.

Alternatively, a real estate agent can get this information for you from the courthouse. Actually, she or he will probably get it from a local title company that also has all the local property records.

Is the Lot Buildable? And Other Burning Questions

Let's say you now have a few likely candidates for your building site. Now what? Of course, the first question you need to ask is this: Is it buildable? You don't want to spend time and money on property that isn't. So here are some important questions to ask about the property before buying it:

- Is the property sufficiently large to site the size of house you expect to build, along with the garage, outbuildings, driveways, and landscaping?
- What are the property's *easements* and *setbacks*?
- If the setbacks are not sufficient to build the home you want, how easy is it to get a *variance*?
- Is the lot zoned for your single-family residence? (Refer to Chapter 7.)
- Does this lot have covenants that are unacceptable to you? (If so, don't try to change them. Just don't build there.)
- Does this specific lot have any other governmental or home-association building restrictions?

- How much privacy does the lot have? Are other lots looking down on it or vice versa?

- If appropriate, how would your home design look compared to neighboring houses? (This is less of an issue on large rural lots.)

- Is the lot in a low-traffic area?

- What's the posted speed limit on the street in front of the lot? What's the actual speed?

- Is the lot on a flood plain (low land, subject to drainage) or a wetlands (seasonal standing water)?

- Is the property and especially the building site easily accessible?

- If crossing another property, is there a legal and recorded easement or right-of-way to the subject parcel?

- Is the view from the lot attractive? If not, can well-planned landscaping solve the problem?

- Will the backyard be sufficiently private? If not, what could you do about it?

- What services does the lot have (power, water, sewer, phone, cable) and which will you need to supply (septic, power, water, cellular, satellite)?

- What are the utility hookup fees for the lot? (They can add many thousands of dollars to the final cost.)

- Are power, phone, and cable utilities aboveground or underground?

- Is there an impact fee required by some local government for the right to join the community? (Really!)

- How much are local property taxes and what do they fund?

- How are property taxes assessed? Size, value, or a formula?

- Are you eligible for any property tax relief due to your age?

- Is the lot convenient to services you frequently need, such as schools and shopping?

- Does the lot have a slope? If so, can you build to ensure that the house's drain line is higher than the public sewer line? (If not, your lot may require a sewer pump at additional cost.)

- What homeowner insurance rates would apply if you build on this site?

- Does water seem to pool in a specific part of the lot without an adequate drainage path?

- Is the soil buildable? Rocky? Sandy? (If there's any question, consider a soil test before buying the lot.)

- Does the lot have good drainage away from the proposed building site and, preferably, toward the road?

- Is the lot in an airport flight path or near an active railroad track?

- Is the lot near a truck route? (Downhill truck routes can be especially noisy if trucks use their "jake" brakes.)

- Is the lot shallow, such as some corner lots, without sufficient depth for an adequate backyard?

- If sewer service isn't provided, will the lot pass a perk (percolation) test and allow a septic system?

- If adding a septic system, is there sufficient room available for the tank, leach field, and, if required by local building regulations, a secondary field?

- If water is not provided, can the site provide sufficient water to support the residence? (A well must produce a minimum gallons-per-minute, or GPM, year-round to qualify for a building permit.)

Building Your Vocab

An **easement** is a privilege of use attached to a specific parcel of land, such as the right of a utility line or a driveway to cross a parcel. A **setback** is the distance from a curb, or other established property line, in which no structures may be built. A **variance** is an exemption to the zoning ordinance for a specific parcel of land.

Code Red

Seemingly quiet streets can become rush-hour raceways if the streets are used to circumvent busy intersections or are a more direct path for subdivision residents. Spend some time at the site to make sure you're not building at the Indy 500.

◆ If off-site water rights are deeded with the property, what is the GPM furnished, and who is responsible for water system maintenance?

◆ Will the lot owner warrant the lot as buildable?

◆ Are there homeowner association dues? If so, are they mandatory?

As you can see, you have many questions that need answers before you even start to plan building on a lot. The most important is the last one: Will the lot owner warrant the lot as buildable? If so, you can move ahead getting answers to other questions knowing that, if there's an issue of buildability, the owner will get the lot back. Even so, it's really up to you to make sure you can build your house there at a reasonable cost.

What's It Cost to Build Here?

So what should you include as you try to figure out how much it will cost to build your house on a specific site? Fortunately, getting answers to the previous questions (and recording those answers in your Home Book) will make the estimating job much easier.

For example, you may need to add the price of a new water well and a septic system to the cost. Or you need to add utility connection fees. Or you may need to remove and replace the existing soil with a kind that's more buildable. Not cheap.

Start with the asking price of the lot. (We'll discuss how to get a lower price later in this chapter.) Add in costs of preparing the lot for a foundation. Not excavating for the actual foundation, but just for contouring the lot so that you can dig the foundation footings. If there's a huge rock that you need to remove, get an estimate of the cost and add it in. Or maybe you need to add or remove soil. Or you need to change a slope and to engineer and provide drainage.

Get estimates on tree removal as necessary. If possible, keep as many trees as you can, especially healthy, mature trees of preferred varieties. Remove brush and junky trees. Also remove trees that will block desirable views or sunlight.

Add in the costs of adding needed services such as power, water, phone, and so on. If it's a suburban lot, include utility hookup costs for that specific lot (some may vary by lot location or size). If it's a rural site, include the cost of drilling a well, installing a pump, and delivering water to the house. Also include the cost of a power system, septic system, and any other services not already at the building site.

Code Red

Be cautious of buying property through a land auction unless you have thoroughly researched the specific parcel. There are typically good reasons why the parcel hasn't otherwise sold. It may not be buildable, or there may be too many restrictions to make building on it economical.

Next, review your answers to the previous questions. Should you include any of those other costs?

Finally, add it all up.

The bottom line is the cost of the building site itself. It's an important number as you compare the cost of building here versus there. You can't compare a suburban lot with a rural site, because one includes all services to the lot line, and the other needs services brought in. Nor can you compare two lots in the same subdivision, or sites in the same area, because each site will have its own challenges to overcome.

Fair Market Price

Okay. You know what it's going to cost before you start building. Now let's consider what it's worth. During your research you have probably considered dozens, maybe hundreds, of potential building sites. Good for you. You've been doing your homework. Along the way you've probably developed a real sense of cost versus value. You can now probably look at the price of a building site and say "too high" or "too cheap; what's the problem with this site?"

Or maybe you can't tell. You need to. Here's why. The book definition of fair market value is "the price at which a buyer is willing to buy and a seller is willing to sell." That's really easy to figure out if you walk into a hardware store and a hammer is priced at $14.95. If you buy it, you are agreeing that it is a fair price. If you don't, you don't agree.

Vacant land is valued in the same way. The seller sets a selling price and potential buyers can say "yes" or "no, thanks." The difference is that, unlike at the hardware store, a buyer can say, "No, but here's what I *will* do …." And, if you can convince the seller that your price is fair to both parties, you get it.

If you're using a real estate agent, the agent can help you determine a fair market price based on comparable sales. But remember that even though the agent is helping you, she or he is actually working for the seller! That's right. It's the seller who pays a commission to an agent from the proceeds of the sale, not you. Not directly anyway. And the commission is based on a percentage of the sale price. So don't expect too much effort from the agent toward getting the rock-bottom lowest price.

Even if you hire a buyer's agent (allowed in many states) you will probably pay the fee based on the final price. Not much help from the agent here either.

It's really up to you to make sure the price is fair and the seller knows that it is fair. You may have to negotiate with the agent before the agent negotiates with the seller. How can you know? An agent can develop an assessment of market value, but it may be biased toward a higher price. Or you can hire a property appraiser to give you an official-looking paper with an expert opinion of the fair market price. It may cost a couple hundred dollars, but it may save you a few thousand.

Back to the definition, though. No matter what any expert says, it can only be a sale if the seller agrees to the price.

Or maybe you're working without an agent. Fine. But the same rules apply, except that you have to be the one to negotiate with the seller to explain why your price is more equally fair than the seller's.

Need some negotiating tips?

- Ask any questions you have about the property, and make sure the seller answers them *before* you make an offer.
- Find out what terms the seller prefers, including an owner contract or a *subordination*.
- Cash talks. If the seller prefers it, offer "cash." Remember that if you finance through a bank, the seller will receive cash. Even ask for a "cash price" to see if you can get an initial discount.
- Before you announce your price to the seller or the agent, build your case for a different price and how you arrived at it.
- If buying without an agent, get a land contract form or one drawn up by an attorney so that the seller can sign right then, if convinced.
- Explain what you want to do with the property, but also explain that you have considered numerous others.
- Don't be in a rush. "I'm considering other offers" typically means "I *wish* I were considering other offers."
- As needed, make sure the land purchase contract includes clauses that will protect you if something unforeseen happens. Will the seller warranty buildability? If so, get it in writing. What happens if it isn't buildable?

Building Your Vocab

Subordination is a change in priority of rights in a parcel of land. For example, the landowner subordinates his rights in land to the house lender in exchange for rights in the land-plus-house.

◆ Remember that many land developers and multiple-lot owners are probably better negotiators than you and/or your agent are. Be prepared with facts.

◆ As needed, mention the primary concerns you've discovered about the property. You may find that they are inaccurate—or you may be able to negotiate a lower price.

◆ Remember that in some deals, getting a seller to agree to required terms (financing, subordination, closing) may be more important than the actual sale price.

◆ Know before you make an offer how a lender will prefer to see you purchase the land. Depending on your credit, down payment, and the local market, the lender may want to pay for the land, or may prefer that you buy it, if you can, as your collateral in the deal. Some lenders are friendlier if you already own the land.

Plunking Down the Money

Congratulations! You just bought yourself a chunk of land! Now what? Someone will whip out a piece of paper for everyone to sign. The seller's agreement will probably be biased toward the seller, just as one you have drawn up will have a bias toward your rights. If an agent is involved, that piece will probably be more balanced, but will include a clause as to who agrees to pay the commission (usually the seller).

On the Level

Negotiate the best deal you can, but have two other prices: your walk-away price and your I'll-take-it price. Don't let a good building site get away for a few thousand dollars that you can afford.

Some sellers prefer that a mutually agreed-upon attorney draw up the papers and that you split the costs. That should be okay unless you have reason to believe that the agreement won't be balanced. If a lender is involved, he or she may furnish an agreement that all review and sign. If so, know about it and have one ready when you get the seller to agree to your terms.

Depending on how the deal is structured, all the paperwork may be handled by the lender's escrow officer or by an independent escrow company. An escrow agent is a disinterested third party who agrees to follow any instructions that the buyer, seller, and lender agree upon. If you all agree that the purchase price is to be paid in shiny new quarters, that's what will happen.

Title and liens are important, too. That's where a title company comes in. For a fee, the title officer makes sure that anybody who has an interest in the land knows about the transaction and is known. In fact, title insurance is sold that protects parties from title loss due to their errors.

What could go wrong? A prior owner or a bank may still hold partial title to the property, and must be paid before the title is transferred. Or someone who has already done work on the property, such as a tree service or a utilities contractor, hasn't been paid yet and has filed a mechanics' lien.

What Happens Now?

You have purchased the land for your new house site. What's next? Lots more paperwork. Chapter 9 will help you to determine a more accurate estimate of the total building costs. Chapter 10 will help you get financing for the whole building package. Part 3 will walk you through getting the needed building permits and following codes, hiring your contractor and/or subs, or deciding to do it yourself and what's involved.

So, don't touch that dial, friend. There's lots more fun coming up!

The Least You Need to Know

- The more you can learn about all available building sites in an area, the more likely you can find the best one for your needs.
- Contact the local tax assessor's or other real estate office to learn about specific properties and ownership.
- Compare the costs of building on each site, as they may vary.
- Negotiate with knowledge about the property, its buildability, and local market conditions.

Chapter 9

Wallet Blues: Estimating Construction Costs

In This Chapter

- Making initial estimates of building costs
- Figuring the costs of materials
- Getting quotes and bids
- Scheduling your home construction
- The right tools for the job

This chapter is where you figure out and total up what your new house is going to cost, which can be a daunting process. This chapter will help you reduce headaches by guiding you through the process of accurately estimating costs. It *really* won't be that bad. And you'll get some help from builders and others.

So pull out your Home Book and let's start figuring out the price of a dream.

How Much Is That Shanty with a Window?

In Chapter 4, I helped you make estimates of what it would cost to build your own home. I also showed you some proven ways of reducing the costs by thousands of dollars. It's now time to get more specific. Why? Because money *does* grow on trees, but the folks who own the orchard, the lenders, want to know exactly how much you need. Just try telling a lender, "Yeah, I think I need about 200 very big ones. Naw, make that 300. Wait, 450 is probably about right."

Even if it's all or mostly your money, it's probably not unlimited. (If it is, call me!) You have an amount you can spend on building your home. You may be able to fudge a little to get the job finished, but you can't go 100 percent over budget, right?

Chapter 4 offered some guidelines for estimating local construction costs based on how unique your house will be. That is, if it's relatively stock and an easier-to-build house, it may be

$100/s.f. in your area. Custom homes with lots of amenities that require special tradespeople can easily be double that figure. A semi-custom home can be anywhere in between.

However, that's not good enough for the lender. And it shouldn't be for you. So let's get more specific.

Making a List, Checking It Twice

If you now have a specific building site purchased or picked out, and a detailed plan of your home, you can get a relatively accurate cost of acquisition and construction.

If you're building this one yourself, or you're at least going to be the managing contractor and hiring subcontractors, you need to develop a *materials take-off*, or list of materials you will need.

Building Your Vocab

A materials take-off is a detailed list of all materials that are needed to build a specific house. It will include everything from foundation rebar to roof fasteners. Quite a long list!

How do you get such a list? Fortunately, a materials take-off can be included in a full plan set. So, if you're buying a stock plan set or hiring an architect, they will probably furnish a materials take-off. If you're designing your own home using professional software, it may be able to produce a materials take-off for you.

Depending on who produces it, the take-off may or may not include everything. That is, it may not include exterior painting materials or driveway concrete or primary landscaping components. Make sure you know what it covers and what it doesn't.

Concrete Foundation Materials			
6 mil Visqueen	1	pcs	10' × 100'
Anchor Bolt, Nut and Washer	40	pcs	½" × 10"
#5 Rebar	84	pcs	20'
6-6-10-10 Remesh	720	sf	
Footing Concrete	10	yds	
Foundation Wall Concrete	17	yds	
Slab Concrete	9	yds	
Optional Block Foundation Materials			
6 mil Visqueen	1	pcs	10' × 100'
Anchor Bolt, Nut and Washer	40	pcs	½" × 10"
#5 Rebar	68	pcs	20'
Footing Concrete	13	yds	
6-6-10-10 Remesh	720	sf	
Slab Concrete	9	yds	
Concrete Block	32	pcs	6" × 16" × 8"
Concrete Block	136	pcs	8" × 16" × 8"
Concrete Block	574	pcs	12" × 16" × 8"
Mortar Mix	24	bags	
Premixed Grout	3	yds	

Basement Framing			
Sill Seal	2	rolls	6" × 100'
Mud Sill (Treated)	8	pcs	2" × 6" × 14'
Exterior Plate	5	pcs	2" × 6" × 14'
Exterior Plate	6	pcs	2" × 6" × 16'
Interior Sole Plate (Treated)	4	pcs	2" × 4" × 14'
Interior Plate	3	pcs	2" × 4" × 14'
Interior Plate	4	pcs	2" × 4" × 16'
Exterior Stud	60	pcs	2" × 6" × 8'
Exterior Post in Wall	1	pcs	6" × 6" × 8'
Interior Stud	58	pcs	2" × 4" × 8'
Steel Column w/Top and Bottom Plates	3	pcs	3" × ½" × 8'
Exterior Header	4	pcs	2" × 12" × 12'
Metal Post Base	1	pcs	6' × 6'

What if your house plans don't include a materials take-off? Well, you can always manually write out the list based on the final plans. Not the easiest route. Another option is to get someone to develop a materials take-off from your house plans. Who? Some drafting services can do this. Or you can take your plans to a building materials supplier and ask them to bid on the materials. They will need to develop a materials take-off first. There you have it.

If you're purchasing a manufactured home, the purchase order to the manufacturer will include a list of things you want in it. It will also include a list of any special requirements, such as matching paint for the site-built garage.

Kit homes have relatively comprehensive material take-offs—or they should. Make sure that yours does. However, it may not be complete. Review it with a builder or sub to see if anything's missing. If so, make a list and get a bid so that you know exactly how much this house will cost to build.

You Can Quote Me!

You don't have to do *all* of the figuring. You'll also be getting quotes and bids from contractors, subs, and suppliers. Even if you're going to do all the work yourself, get bids on construction so you know what it costs—and you'll save. Who knows? You may eventually decide to hire someone to tackle the tougher jobs.

So how can you get bids on constructing your house? Ask for them! You can call general contractors and/or subcontractors in your area to bid on construction (see Chapter 12). Where can you find them? First, check the telephone books for the area in which you're building. (Contractors charge more if they have to drive a long way to the building site.) Check the business section under these and related categories:

◆ Building Contractors

◆ General Contractors

◆ Demolition Contractors

Ka-ching!

Most builders add in a fudge factor. It's typically an additional 10 percent to the cost of materials to compensate for things forgotten on the take-off. If it doesn't look like you'll need it, use this extra to upgrade materials.

- ◆ Excavating Contractors
- ◆ Concrete Contractors
- ◆ Foundation Contractors
- ◆ Framing Contractors
- ◆ Electrical Contractors
- ◆ Plumbing Contractors
- ◆ Dry Wall Contractors
- ◆ Plastering Contractors
- ◆ Heating and Ventilating Contractors
- ◆ Air Conditioning Contractors
- ◆ Insulation Contractors
- ◆ Masonry Contractors
- ◆ Flooring and Floor Covering Contractors
- ◆ Paving Contractors
- ◆ Water Well Drilling and Pump Contractors
- ◆ Sewer Contractors
- ◆ Landscape Contractors
- ◆ Fencing Contractors

Get estimates and bids even if you plan to do the work yourself. You'll know what you're saving, and you may decide that you'll save the money elsewhere and let someone else do the dirty work.

Whether you're doing your own building and estimating or using contractors, you want to know what you're paying for and whether the estimates are accurate. So, let's look at the components of the full building estimate. It will help you to figure out how much of it you want to do as well as tell you if a contractor is taking advantage of you.

Building Your Vocab

Loaders push and lift dirt to remove it and contour the land. **Backhoes** pull and lift dirt to dig trenches for underground utilities. Some units have a loader on one end and a backhoe on the other.

Digging In to Excavation

Excavation of a lot prepares it for the foundation and driveway. The work is done by heavy equipment called *loaders* and *backhoes*.

Excavation contractors charge by the hour. Based on experience, they can estimate the time required to give you a bid. However, the bid will probably include a fudge factor.

Depending on the property, preparation, excavating the foundation, and grading the building site will take one or two days. The excavator may first need to cut down or have someone else cut down trees, so you may need to factor in this service as well.

A Good Foundation

Most houses are built with concrete foundations, or at least concrete footing onto which blocks or other masonry sits as the foundation. Forms for the concrete must first be built. Foundation contractors will charge by the hour or by the foundation's square footage.

Concrete is purchased by the cubic yard. A cubic yard of concrete will give you 27 cubic feet (c.f.) or about 46,656 cubic inches (c.i.). A cubic yard will give you about 80 s.f. of concrete 4" thick. That's a 20' × 40' × 4" slab. Or 30 lineal feet of 8" × 16" footing. Or about 40 s.f. of 12" × 12" × 8" poured concrete walls.

Tables are available from concrete suppliers to help you make a closer estimate of how much concrete you'll need for your building site. With a table you can plan it yourself, or make sure your foundation or concrete contractor is estimating accurately. Doesn't hurt to double-check.

You're Being Framed

Residential construction framing is bid and quoted in different ways. The bottom line to the contractor is estimating how many hours the job will take and multiplying that by the contractor's hourly rate. The quote will probably be given as per-square-foot, based on the contractor's experience.

Depending on who is doing what, the quote or bid may be broken down as floor framing, wall framing, and roof framing. The roof framing may include installation of pre-engineered roof trusses.

Code Red

Remember that special rooflines and dormers or gables will cost more than a standard roof. Of course, they can add more room on the inside and more visual value on the outside.

On the Side

The framing contractor may do siding, if wood, or a masonry contractor may do it, if brick or rock. Or both may. They will price estimates and bids by the hour, by the square footage, or by the sheet. Some will estimate by figuring how many full 4' × 8' sheets they will need and how many cut sheets (doors, windows, gable ends). They will add for additional trim that's needed such as corner trim boards.

Installing lap siding is more time consuming, so expect costs to be higher than sheet siding. There will be more waste, too. Masonry siding is even more labor intensive and expensive. You may need a separate bid from a masonry contractor.

Walls and Ceilings

In most locations, insulation requires one specialized contractor, and drywall requires another. In smaller communities and rural areas, one contractor may do both—and frame the house.

A contractor typically bids insulation by the type and volume. Blown-in insulation requires special equipment and knowledge. Blanket and batt insulation is easier to install. (See Chapter 6 for a refresher on types of insulation.)

Contractors bid drywall by the size of the wall and ceiling surfaces. They might make the estimate using the house's square footage, but they should make a final bid on the drywall surface, to be more precise. Plans with lots of closets, bathrooms, and other smaller rooms, or with nonstandard ceilings, will cost more to drywall because they take more labor. If you're buying the materials yourself, plan for about 10 percent waste.

Finishing Up

Like many house components, contractors typically bid flooring based on square footage, but they actually calculate it on the estimated hours it will take to prepare and install.

Carpet is the fastest and easiest to install, as it comes in large sheets that the workers join and cut as needed to fit the room. They install tack strips around the room perimeter, and lay down padding first. They will probably base bids on square footage, but calculate them on an hourly work-rate. Stairs are bid at a different rate, as they are more labor intensive.

Contractors will also bid wood, vinyl, laminate, ceramic, and other types of hard flooring by the square foot. Obviously, the more difficult the installation, the higher the s.f. installation bid. What gets really expensive is cutting individual ceramic tile for mosaics or borders.

Other Stuff

Estimates and bids for other house components work the same way. Most are calculated at the contractor's hourly rate, but quoted at a per-size or per-unit rate, based on experience. The experience is either that of the tradesperson or a trade organization that publishes bidding guidelines.

On the Level

Bids are typically lower than estimates, unless something changes. That is, a final bid for plumbing will probably come in at or a little less than an initial estimate from the same contractor, as long as no new fixtures have been added. Estimates are ballpark figures … and some contractors are hoping for a home run!

For example, a septic system contractor will bid based on the tank size, the length of field lines, the distance to the house, and other factors. An electrician will bid by the receptacle, adding in extra costs for hooking up built-in appliances and other labor-intensive work. A plumber will charge by the fixture installed or by the hour. You get the picture.

What Am I to Bid?

Now that you know what bids you need, it's time to start collecting and analyzing them. Remember that both you and the lender will want *written* bids. Typically, a bid will tell you what the contractor or sub will do, maybe even when they will do it, and how much it will cost. In fact, most bids are signed and, once you sign, are legal contracts.

The first step in calling for bids is identifying whom you want to bid on the various jobs in the project. Your general contractor will do that, or you will if you're the builder.

Next, ask contractors what they need to bid. Besides the standard stuff, the contractor may indicate other things needed to make the bid. Or the contractor may not be taking on any new work for a few months. Things you need to know.

Then, gather the specifications the contractor needs to know about the project and what to bid on. For example, a potential framing contractor will need a full set of plans, the house specs, materials take-offs, who is supplying what materials, when the work needs to be done, and to whom to submit the bid. Include a reasonable deadline for accepting bids. Remember to keep track of the bid requests you send out so you can jog someone's memory before the deadline.

Ka-ching!

Are you your own GC? If so, look for a lender with owner-builder experience. He or she can help you call for, analyze, and award bids.

Finally, accept the best bid(s). That doesn't always mean the lowest bid, however. Sometimes a low bidder is such because they plan to undercut material requirements. Or it may be because the other bidders are quite busy and you'll need to pay a premium to get them. Of course, you can still negotiate price and scheduling once you have all the bids in.

Should you decide now which contractors and subs to use? Not yet. You still need to talk with Ms. Moneybags, the lender, coming up in the next chapter.

Scheduling for Success

There's lots of work to do when building your own home. That's evident. What may not be as evident is how important it can be for someone to stay on top of the building schedule. If you've hired a general contractor, he or she will handle scheduling. If you're contracting your house, the person staying on top of the schedule is you. Even if you're building the whole darn thing yourself, you need to schedule delivery of materials and the building department inspections.

If you're good at coordinating multiple tasks by multiple people, life will be easier—but not easy. If you're not so good at scheduling, consider hiring someone. Or at least learn how to manage large projects. It's in the details. And there are lots of details to building your own home, no matter who actually does the construction.

First, let's take a look at the 20 major processes needed to build a single-family residence on a standard lot with utilities at the curb:

- Preconstruction preparation
- Excavation
- Pest control
- Concrete
- Waterproofing
- Framing
- Roofing
- Plumbing
- Electrical
- HVAC
- Siding and/or masonry
- Barriers (doors and windows)
- Insulation
- Drywall
- Trim
- Painting
- Cabinetry
- Flooring and tile
- Gutters and downspouts
- Landscaping and driveways

Most construction problems are solvable. It may mean paying a premium to get the drywaller to push his schedule, or threatening to flatten the tires on his SUV. (On second thought, maybe you can just turn on your inherent charm.) The real challenge of scheduling is what's called managing the critical path. For example, the drywall sub is only available next Tuesday, but the electrician and insulation contractor can't make it until next Thursday. You can't have drywall installed before the interior of the wall is finished. "Hey, contractor. That's *your* problem." Except if you are your own contractor. Fortunately, you'll learn more about what is critical path and what isn't as you read this book. And I'll give you tips along the way on how to keep things moving toward the goalpost.

The Right Tools

If you do any of the work yourself, you're going to need the right tools for the job. Which tools? It depends on the job.

Here's a list of the basic tools needed to build a house. Many tradespeople have additional specialized tools:

- Lots of carpenter (flat) pencils
- Tape measures (25' and 100')

- Carpenter's or combination square
- Chalk line and plumb bob
- Levels (2' and 4')
- Framing and finish hammers
- Power nailer
- Pry bar and nail claw
- Sledge hammer
- Utility knives
- Power circular saw
- Power sabre saw
- Power miter saw (or miter box and saw)
- Power table saw and/or radial arm saw
- Screwdrivers
- Power screwdriver
- Power drill and bits
- Extension cords (10- or 12-gauge)
- Air compressor (if needed by power tools)
- Adjustable wrenches and pliers
- Files and chisels
- Nail sets and countersinks
- Finishing sander
- Ladders

As I mentioned, many other tools are used on construction sites for special purposes. To give you a hint of what they are, here's a partial list:

- Builder's transit (or laser level)
- Hardwood flooring nailer
- Power finish sander
- Wire strippers
- Plumber's wrenches
- Aviation snips
- Ripping bar
- Reciprocating saw
- Pneumatic stapler
- Volt-ohmmeter
- Drywall taper
- Paint brushes, rollers, and trays
- Paint sprayer

Should you buy these tools now? No. If you're just building one house, buying expensive equipment may not be practical, especially if you might end up hiring someone who has all the tools they need—and even better ones. However, the next time you're at the local building supply store (and you will be there *many, many* times!) start looking at tools. Ask clerks for recommendations on buying the best framing hammer or table saw. Get ideas. Get opinions.

If you don't want to buy all the tools, one option is to rent what you need. A good rule of thumb is to consider renting anything that costs over $100. Also consider renting anything that you don't need, but that will save you lots of time (such as a power nailer).

Another option is to borrow. I know, "Never a borrower nor a lender be." Like anybody really follows that advice. As you start to talk about building your own home, you're going to meet folks who have been there, done that. They may even have some tools they no longer need. Maybe you can borrow them, or at least buy them for less than what they would cost at your building supply store.

Ka-ching!

To get the most from your tools, you must take care of them. That means, for example, not leaving them out in the weather, keeping cutting edges sharp, and repairing or replacing frayed electrical cords. With proper care they will last longer and work safer.

In addition, large swap meets and community garage sales typically have at least a few booths loaded down with contractor tools. But be careful about quality. Some flea-marketers import cheap tools for resale. Don't even try to pound a nail with a $3 hammer!

You're getting there. In just a few chapters, you've designed an energy-efficient house, selected a prime location and building site, and calculated what it will cost to build your own home. The next major hurdle is talking with the lender, coming up in Chapter 10.

The Least You Need to Know

- Your initial estimate can help you determine how much of a house you'll get for your money.
- The materials take-off is a comprehensive list of all materials needed to build a specific house.
- Contractors, subcontractors, and tradespeople will quote or bid on your project if you give them specific plans.
- Smart scheduling can save you thousands of dollars and lots of time and frustration.
- Make sure you use the right tool for the job—even if you have to rent it.

Chapter 10

Financing Your New Home

In This Chapter

- What are your assets and liabilities?
- Meeting your lenders
- Applying for a successful construction loan
- Closing the loan

Grab your wallet! It's time to find out how you're going to actually pay for this dream home. Sure, you made some guesstimates in Chapter 4. Now you'll find out exactly what you can afford and how you can pay for it.

This chapter offers a real-world look at lenders, the lending process, construction loans, mortgages, and closings. Just as important, it offers proven techniques for saving real money on financing your own home.

As before, pull out your Home Book for note taking, and let's see how you can get the most for the least.

Anything to Declare? Assets and Liabilities

First, let's talk about assets. The home you're planning to build is an asset. In fact, anything that has value is an asset. If it has value to folks other than just you, you can set a monetary price on its value. Whether you ever plan to sell your house, an asset, isn't important right now. You might someday, or your grandchildren might. So the market value of your house is important. (Remember, market value is the price at which a buyer is willing to buy and a seller willing to sell a specified asset.)

And we might as well talk about liabilities, too. These are your debts. If you finance your home with a construction loan and, eventually, a mortgage, these are liabilities. You promise to pay someone later for what they give you now.

Finally, what you would have left if you sold all your assets and paid off your liabilities is your net worth. You sell an existing house for $200,000 and retire the mortgage of $120,000, leaving you $80,000 in net worth. That makes sense.

So what?

Lenders are going to want to know your net worth. If you sold all your assets today and paid off all your debts, how much money would you have left?

Fortunately, lenders also consider jobs and education as assets, though intangible. It's earning power. You may not have a dollar of net worth, but if you're in a good job and have a trade that will keep you employed for a while, you can pay some kind of a mortgage. That's what they're interested in.

Having said all that, what assets do you have that can be used to help you finance the building of your own home? Let's take inventory. In your Home Book, start listing and assigning a value to your assets:

- House, condo, or vacation home you now live in and own or make payments on
- Vacant land you now own or pay on, including the land you want to build on
- Vehicles (cars, trucks, boats, aircraft), paid for or not
- Stocks, bonds, mutual funds, pensions, annuities, and other financial investments
- Businesses you own, partnership, or hold stock in
- Bank, savings, credit union, and other liquid asset accounts
- Income from rental properties
- Personal loans owed to you
- Alimony and/or child support income
- Household furnishings and home computers
- Collectibles (if significant in value)
- A solid job with probable future employment
- A trade that will keep you employed for many years
- Estates you expect to soon inherit
- Winning PowerBall lottery ticket in your dresser

You're listing these assets now because the lender will want them later as you fill out your loan application.

Back to liabilities. List them in your Home Book as well:

- Current mortgages and land contracts
- Current rent, if any
- Vehicle loans
- Credit card balances
- Personal loans you are paying back
- Debts against stocks and business investments
- Alimony, separate maintenance, and child-support payments
- Pledged assets (things you own but plan to give to someone else)
- Your promise of splitting that winning PowerBall lottery ticket with your ex

On the Level

Most mortgage lenders don't like to finance bare land. So buy your lot and make some payments to get equity built up in it before you talk to them. The lot owner may offer short-term financing until your permanent mortgage pays the balance. Of course, the more equity you have, the easier it will be to get a construction loan.

Selling Your House

Do you need to sell your present home (if owned) before you build your new house? Probably. Unless you have lots of money and other assets, the lender will want you to sell your current house and pay off the mortgage before you take on another mortgage.

Most folks who are building their own house can't afford two mortgages. What do they do? They typically sell their current house and move into a rental while the house is being constructed. The strapped-for-cash ones may even park a trailer on the building lot. A few even go live with relatives—not recommended for the faint of heart. (My parents lived with us for eight months while their home was being constructed!)

Selling Other Valuable Stuff

Depending on what you own, what you need, and how easy it is to sell the assets, you might consider selling some of your other stuff before you go face the lender, such as unnecessary vehicles, unprofitable stocks and other investments, and collectibles. In addition, consider building your cash funds by doing the following:

- Selling annuities for cash
- Discount outstanding loans for cash
- Factor (sell) bad debts
- Consolidate loans and credit card balances
- Cash in that winning PowerBall ticket!

The point to this exercise is to not only increase your liquid assets, but also to lower your liabilities—or at least consolidate them so that your monthly payments are less. Exactly what you do to make this happen depends on your net worth.

Understanding Lenders

Here's a secret: lenders *really* don't want to turn down your loan! In fact, lenders *only* make money if they make loans to people who pay them back. That makes good sense. So help them! Help your lender discover that you are a good credit risk and that the investment made in building your own house will pay off.

> **Ka-ching!**
> Every $1,000 less you have to borrow will save you about $3,500 over the life of a typical 30-year mortgage!

How can you help your lender? First, by knowing what they need to make the lending decision. The previous section of this chapter helped you figure your net worth (assets less liabilities). That's important. So is their knowing that your work history and future are solid. As solid as anyone's! No one expects to be at the same job for the life of a 30-year mortgage. So the lender wants to be sure you have a trade or at least some valuable skills that will enable you to pay the mortgage payment each month.

It's clarification time. What you'll first be going for is a short-term construction loan. However, it will be replaced by a long-term mortgage, probably by the same lender. In fact, the lender won't want to touch a construction loan unless a mortgage will then pay it off.

Let's talk about construction loans for a moment. As I mentioned, a construction loan is a short-term loan to finance the construction of a house. You may build the entire house yourself, participate as the general contractor or a subcontractor, or simply be a knowledgeable owner who will keep everyone honest.

> **On the Level**
> Think you have some credit issues? Don't be shy; ask! Get a copy of your credit report from a lender and go over it for advice on how to improve your credit. You may learn that a construction loan isn't likely this year, but work at it for a while and you'll get one.

The construction loan won't be paid out all at once. The lender will want to make sure that the construction is progressing as planned. So the lender (with the contractor) will establish milestones where specific tasks are done and can be paid. The payments are called draws because money is drawn out of the construction account.

Who actually makes sure the work is done and then requests and receives the draw depends on how the construction loan was, er, constructed. Here's a common draw schedule:

1. Twenty percent when footings, foundation, first-floor joists, and subfloor (or slab and plumbing rough-ins) are completed.

2. Twenty percent when roof and interior are sheathed and all interior partitions are roughed-in.

3. Twenty percent when plumbing, electrical, HVAC, insulation, drywall, siding, windows, doors, and roofing are installed.

4. Twenty percent when interior and exterior trim, stairs, casework, and countertops are installed.

5. Twenty percent when fixtures, plumbing, electrical, lighting, flooring, appliances, driveway, and landscaping are installed and all interior and exterior surfaces are painted.

Of course, there are lots of variables here. "Drywall" might be stucco or finished logs. A basement might be in there somewhere. An attached garage is typically included. In addition, the lender might not draw up the final payment until he or she knows that everyone has been paid up to that point. Contractors, subcontractors, and suppliers will have to sign a Release of Liens first.

As you can see, one of the general contractor's jobs is to make sure that work (and materials) don't get too far ahead of payment. That is, the roofers will want their money as soon as they're done (draw 3). They don't really care if the lender has only paid draw 1. You'll learn more about what it takes to be your own general contractor in Chapter 13. I promise!

How Loans Are Processed

Okay. You now know how the construction loan works. Let's see how to get one. First, understand that we live in a world of specialization. That is, some folks do one thing very well (and other things poorly). Lenders are the same way. Don't even ask your credit card company to finance your new house. They don't have a clue. And they don't *want* to know.

So the first step in getting a good construction loan is to begin looking for a lender who specializes in them. Professional builders typically know where to go. If you're hiring a general contractor to manage construction or to help you do so, ask about lenders. The lender doesn't have to be local, but it's best to use one that knows the local building conditions.

In addition, consider your own banker or savings and loan (S&L). S&Ls typically like making mortgages and banks prefer business and consumer loans, but the difference between the two is diminishing. Credit unions prefer automobile loans, but have been known to tackle mortgages and even construction loans. Knowing the lender helps.

On the Level

Fannie Mae is the nation's largest source of financing for home mortgages. They can handle single-family residential mortgages up to about $333,000 without breaking a sweat.

The loan process will begin with an application. Your lender may have its own form or it may use the four-page *Uniform Residential Loan Application*, or URLA. It's also called the Fannie Mae Form 1003 or Freddie Mac Form 65. Even if your lender uses another form, most will accept this one. Just make sure that the URLA form you use is the latest version.

Why do these two people, Fannie Mae and Freddie Mac, care about loan applications? Good question. Actually, Fannie Mae is a nickname for the Federal National Mortgage Association (FNMA). It's a publicly owned, government-sponsored corporation that buys mortgages from lenders and resells them to investors. The Federal Housing Administration, the FHA, backs many of the mortgages. (I don't think they have a cute nickname!)

Freddie Mac is a nickname for the Federal Home Loan Mortgage Corporation (FHLMC). It's a publicly chartered organization that bunches mortgages into investment packages. It's owned by savings institutions and watched over by the Federal Home Loan Bank System.

Fannie and Freddie have standardized the home loan process to make it safer for lenders and for investors. That means if your loan application meets these standards, chances are you'll get the loan. Much more predictable than basing the decision on whether or not the lender had a fight with her husband that morning!

And if you're a veteran, there's good news. The Veterans Administration (VA) offers loan assistance as well. Actually, the VA helps package the loan and insures it, making it more attractive to investors. Not to be outdone, the FHA has special loan opportunities as well. Your lender can tell you more about them. (They change!) Building in the country? The Farmer's Home Administration (FmHA, with a lowercase "m" so that it's not confused with the FHA) is part of the U.S. Department of Agriculture. It runs programs to help people buying and building homes and farms in small towns and rural areas. You'll hear it called "Farm Home."

Bring the following paperwork with you when you first visit the lender:

◆ Completed URLA or the lender's preferred application

◆ Bank and credit card statements with account numbers and recent balances

◆ Current pay stub and employer information for all those who will be signing the loan

◆ Investment, pension, and life insurance valuations

◆ Last year's tax returns and probably the prior year's as well

◆ Social Security numbers for those signing the loan

◆ Call ahead and find out if there's anything else you should bring—or plan on dropping it by later the same day

So, what's the lender going to be looking for on your application? Of course, each lender will have specific criteria, and most won't help you out by telling you *exactly* what they look at. However, most follow some guidelines. They are commonsense rules developed over years of lending experience. Stuff like this:

◆ **Mortgage payment-to-income ratio (MR).** Your monthly mortgage payment, including principal, interest, taxes, and insurance (PITI), should be about 29 percent of your *gross* (total) monthly income.

◆ **Total debt-to-income ratio (DR).** The total amount of your monthly debt payments, including the new mortgage, car payments, and credit card payments, should be less than 40 percent of your *gross* monthly income. (Some want lower DR, others will go a shade higher for applicants with great credit.)

◆ **Loan-to-value ratio (LV).** The ratio of the total appraised value of the house and land to the loan amount. Most conventional loans have an LV of 90 percent or less, and the loan is no more than 90 percent of the home's appraised value. Some loan programs will go higher, but tack on private mortgage insurance (PMI) and up the costs.

Again, we're talking about home mortgages rather than construction loans, but one will lead to the next, so these guidelines typically apply to both.

Let's look at it from the lender's perspective. You're the lender, and these nice folks have come to you for a new construction loan and, hopefully, a subsequent home mortgage. You've run a credit check and it looks good. They seem like fine people who pay their debts. That's important. How do you make your decision?

 Code Red

Even if you're using a general contractor, you will need to get the construction loan yourself. The GC may help, but you're the one who will pay off the loan, so make sure you're comfortable with it.

First, you calculate the MR (mortgage payment-to-income ratio) to figure out the maximum they can pay each month for their house. Then you look at their other debts (DR, debt-to-income ratio) to make sure these fine folks don't have more debt than they can reasonably pay. If their DR is high, you might counsel them on how to consolidate or lower their DR so *you* can loan them more money!

As a smart lender, you know there are other things that can be done to help these really nice people buy a loan from you. You know that the ratios may be slightly different for fixed-rate, adjustable, and graduated-payment mortgages. Yes, you're only dealing with a construction loan at this point, but one that will probably be converted into a mortgage down the road. So, you apply these other ratios to see what works best.

You may also advise them that much hinges on the lender's appraisal of the finished home (called a *determination of reasonable value*). What's it worth?

You won't confuse these folks with an explanation of the various types of appraisals. You know that a lender's appraisal is typically more conservative than an appraisal of market value. Hey, lenders are conservative by nature!

So you counsel them that the more value that can be added to the house without adding cost, the higher the final appraisal will be—and the higher their construction loan and mortgage can be. Here's what you tell them will add value:

- Adding labor and materials to the house (sweat equity) that don't add equal cost will increase value.
- If qualified, managing the construction of your house can save lots of money and reduce the LV ratio, making it a safer loan (for you, the lender!).
- Add amenities in the design that translate to salability without adding lots of extra cost.
- Build on a premium lot with a location or view that will appreciate faster than nearby lots.
- Shop around. By hiring good contractors and buying quality materials at the lowest available costs, you add value.
- DON'T OVERBUILD! A house that is worth 20 percent more than the neighbors' homes will *increase the value of the neighbors' homes*. It won't help the owner as much. Build appropriately for the location, or even a little under the typical value, so that other homes can add value to yours.

Negotiating the Best Loan

You're a smart consumer, so you're not applying to just one lender. You have two or three who are considering your application. How will you consider the offer from each of the lenders? First, know what the *total* costs will be for the construction loan—and start asking about the terms of the final mortgage.

Ka-ching!

Many construction loans include clauses to automatically roll over to a home mortgage once the occupancy permit is issued. Don't wait until then to begin negotiating the mortgage. Start now.

Total costs include the following:

- Loan origination fees
- The construction loan amount
- Recording fees, if appropriate to your area
- Credit report fees
- Escrow fees
- Mortgage discount fees, called "points"
- Appraisal fees
- Title search and insurance fees

- Mortgage insurance
- Construction site insurance (if you're not using an insured GC)
- Prorated property taxes (in some areas, it takes the local tax assessor a year or more to get your new house valuated and on the tax rolls, so your first year's property taxes may be the value of the empty lot—or not!)

Some of these fees are nonnegotiable. What it is, it is. However, others are negotiable. Or at least you have more options than those initially offered. For example, you may be able to negotiate a lower loan origination fee or make sure it covers both the construction loan and the ultimate home mortgage.

Also, some lenders will absorb more of your fees than they initially offer. Ask for them to pay the escrow fees and recording fees, or lower the appraisal fee. You can often reduce mortgage points through negotiation. It's worth a try.

You'll also be able to negotiate whether the construction loan is converted into a *fixed-rate* or *adjustable-rate* mortgage (ARM). The difference can impact your initial mortgage costs and payments.

> **Building Your Vocab**
>
> A **fixed-rate** mortgage is a loan with an unchanging interest rate throughout its life or term. An **adjustable-rate** mortgage (ARM) is a loan in which the rates of interest and payment change periodically, based on a standard rate index. Generally, the ARM has a cap on how much the interest rate may increase.

That's a Wrap!

Congratulations! Through hard work and perseverance you've found an experienced lender, presented your application, negotiated a fair construction loan, and been approved for funding. Now it's time to close. Closing the loan means sitting at a table with a closing or escrow officer and going over a big stack of paperwork, signing as indicated.

Hopefully, you've gotten the answers you need well before it's time to sign papers. But if you haven't, don't be shy. Ask. If you have a GC, make sure she or he is there to help you through the process. Also ask that the loan officer who took your application be there in case you have questions. Don't be shy about asking your questions, and make sure you understand the answers. You're paying for them. It's very important to make sure you understand the draw schedule for your construction loan. It is your road map with all the gas stations along the route clearly marked. Miss one and you'll be walking! Once everything is signed, the lender will set up an account from which draws will be paid as specific conditions on the draw schedule are met.

Let's par-tee! You've designed your home, found a great location, and figured out how you're going to pay for it. Good going! Now, all that's left is to build the darn thing. With help, of course.

The Least You Need to Know

- Count up your assets and liabilities before going to visit a lender.
- Knowing what a lender wants can help you get what you want.
- Loans are commodities, so shopping around can save you thousands of dollars.
- Make sure you understand the loan's terms, especially the draw schedule and how it will be paid.

Part 3

Hiring Some Help

You're going to need lots of help as you build your dream home. You may need a contractor or an advisor, subcontractors, an architect, an engineer, and material suppliers. Even if you're buying a manufactured home, you'll need help from a few of these folks before moving in. Or maybe you're going to do most of it yourself. You'll at least need to buy materials.

How and where can you find these helpful folks? And how can you keep them from helping themselves? In this part, I offer answers to these questions. Reading these chapters can also give you the confidence you need to take on more of this project than you initially planned.

Hey, you can do this! Every year, more than 100,000 owners build all or part of their own homes. Yours will soon be among them!

11

Codes, Permits, and Legal Stuff

In This Chapter

◆ Learning about local building codes
◆ What permits do you need to build your house?
◆ Understanding building inspections
◆ All about contracts, deeds, and liens

It's said that a job can't be started until the paperwork weighs as much as the product. Fortunately, that's not true of houses—but you may soon think so as you gather plans, get estimates, read codes, submit forms, sign papers, resubmit forms, sign more papers … well, you get the idea.

In this chapter, I'll guide you through the labyrinth of codes, permits, contracts, and other paperwork that seemingly intends to make your life more difficult. The chapter assumes that you will participate in the permit process. Even if you're going to leave it up to a general contractor or advisor, you'll want to know what's involved.

Just remember that most construction paperwork has an important function. Codes standardize. Permits enhance health and safety. Legal documents keep everyone relatively honest. At least that's their intent.

So before you start pouring concrete or pounding nails, read over this chapter, Home Book in hand, to find out what you must know about the paperwork side of building your own home.

Deciphering Codes

You've heard about building codes. They are the standards set up by someone somewhere to seemingly make it more difficult to build your house the way you want. Actually, the codes are designed for the health and safety of you, the residents of your house, neighbors, and anyone who buys the house in the future.

So does every county, city, town, and burg have its own building code? Kind of. Most adopt a standard building code with maybe some modification, depending on local needs and building practices. Building codes were developed as reactions to fires, earthquakes, hurricanes, tornadoes, and other disasters to minimize future damage to buildings. That means the codes required in earthquake-epicenter Los Angeles will certainly be different than those in rural Montana.

National building codes recently have been standardized. The western United States previously used the Uniform Building Code, or UBC; the southern states followed the Standard Building Code, or SBC; and other states used the National Building Code, or NBC. They are all now rolled into the International Code (www.intlcode.org) to make construction more consistent nationally.

Getting Your Building Permits

Where can you get a copy of the local building codes? From the building department that has jurisdiction over the building site. It's typically a county, city, or town building department, so check the local phonebook under Government. Tell them where you're planning to build, and they can tell you whether or not they have jurisdiction and how to get a copy of the building codes.

Your building plans, drawings, and specifications must reflect the appropriate building codes. So, whoever draws up the plans must know them. Also ask the building department if the jurisdiction has a master plan or guidelines that you must consider when building. Keep in mind that you may have to answer to more than one governmental body when building your house. For example, a permit in a township may require approval by the township's architectural review committee before or during submission to the county that actually makes the decision.

If you're buying your house plans from a plan service, make sure the plans fit the requirements of local building codes, or can easily be modified to conform, before paying for them. Depending on how stringent the local building-permit process is, consider having a local architect or draftsperson review and revise the plans to meet local requirements.

So what plans and paperwork are you going to need before you can apply for a building permit?

Here's a typical list:

- **Plot plan** of the entire parcel with all existing and proposed structures
- **Floor plan** with the location, size, and use of each room, location and size of windows and doors, location of electrical outlets and subpanels, and location of plumbing and heating fixtures
- **Foundation plan** with all dimensions including exterior and interior footings, stem walls, pier blocks, and foundation support, and including footing depths, rebar, and anchor bolt locations
- **Elevation plan** of the finished exterior including all openings, siding material, original and finished grade, roof pitch, and roofing materials
- **Framing plan** for floors and roof including lumber grade, floor girder size and spacing, floor joists, wall studs, ceiling joists, and roof rafters and/or trusses
- **Cross-section plan** showing all primary structural elements from the foundation to the roof including heights and clearances
- **Signatures** of plan designers and engineers as required by code
- **Other stuff** including structural and engineering calculations, soil reports, and permits by other agencies as required by code and local building officials

Ka-ching!

Ask your building department if they offer a preliminary plan check. If so, they will review your building plans and note anything they think will keep them from being approved. You will typically be charged a fee, but you may be able to set up an appointment and get everything reviewed on the spot. It's really a time- and money-saver.

Code Red _____

Building departments won't decide on your permit until *everything* they need is in the package. That may mean first getting reviews and approvals from a planning board, development commission, environmental health, air quality, public works, fire department, and other agencies as required. Make sure you know what they need and how to get it before submitting your package. In some areas the filings are concurrent, while in others they must be consecutive in a specified order. Some will check with the other authorities for you as part of the process.

Of course, you'll have even more plans than what you give to the building permit department. You may also have additional elevations, door and window schedules, room finish schedules, detail sheets, and a full HVAC schematic. These plans and papers will be more important to the contractors than to the building department.

First, you'll complete a Building Permit Application form and submit it with all the stuff to the appropriate building department. In most cases, you'll need a specified number of copies. Don't expect to file and walk out with a building permit the same day. Ask in advance, but the building permit process typically takes one to six months, with three months as an average. Much depends on what is required and whether they have a backlog of permits under consideration.

Of course, you will be charged fees for the time and effort to consider your building plans. Some fees may be due when you file the application. Others may be due when the permit is issued. You may also need to pay any special levies, such as fire or school district, at that time. In addition, many locations require a utility hookup fee. Your local building department will help you figure out who wants what.

Next, the building department staff will review all the plans for compliance to requirements. Some departments will call and ask questions they have. Others will just shoot the plans back to you and say "no" or "start over."

The staff reviewer will make a recommendation to the department manager or whomever else has the power to say yes or no. Unless the house is controversial (such as lots of neighbors complaining), the manager will follow the staff's recommendations (so be *nice* to the staff!).

You've got your permit.

Okay. What you've been seeing is the building permit process for a "typical" house in a "typical" community. Now let's talk about some of the atypical components that the local building department will help you resolve:

- An on-site septic system will probably require additional permits from local officials.

- An on-site water well or spring and delivery system will also need special dispensations from the utility department.

- Trying to build in a higher-use zone (a house in a commercial zone, for example) may need to be planned and permitted differently along with a special dispensation from the local planning commission.

- Building in a designated development area (coastal, for example) may require that plans go to a special review board first or simultaneously.

- Building a unique design or system (log home, post-and-beam, and so on) may require educating local officials on industry standards and on how they should review the plans.

Ka-ching! _____

Don't want to mess with the building permit process? Hire someone to do it. Most larger municipalities have permit expediting services that check plans, review corrections, submit plans, and obtain permits for you for a fee. Check your local phone book under License Services and Construction Consultants. Also check Appendix B for online permit resources.

- Building close to water may require an environmental impact filing.

- Including an in-ground pool or a fuel-storage tank may require special plans, permits, and paperwork.

- Attaching a site-built garage to a manufactured home may require a special permit from the state department for manufactured housing.

- Utilities that require the installation of a new power pole or underground utilities run may need a special permit or work from the local utility company.

The building department will not go away once you have a permit. In fact, it is your partner as you build. To make sure everything goes as planned, a building inspector will periodically visit the site to make sure that everyone is meeting standards.

Here are typical inspection points for a single-family residence construction:

- **Foundation inspection** after footing excavations are done, forms are built, and rebar is in place (depending on the type of foundation).

- **Slab or underfloor inspection** after the concrete slab, if any, or the subfloor is installed.

- **Frame inspection** after all framing, bracing, and roof sheathing are installed and required utility rough-ins are done.

- **Wall inspection** after exterior sheathing and drywall or lathing is installed, but before plaster or tape is installed.

- **Final inspection** after the building is completed, the lot finish-graded, and the house is ready for occupancy.

- **Other inspections** may be required by the building department depending on the type of construction and the plan's complexity; the building department will notify you in advance of additional inspections.

On the Level

The installation and preparation of my parents' manufactured home required that they install a new power pole. Electricity was needed during construction. The bad news was that the utility had a three-month backlog of poles to be set, no exceptions. Hurry up and wait!

In most cases, inspections must be "called for" prior to the inspector visiting the site. Find out how much advance notice is needed. Some will be able to inspect the same day while others may require a week, depending on how busy the inspectors are (and how much clout the builder has!).

The Legal Side of Building

Whew! That was hard work, but very important. Once you have your building permit you can begin, in earnest, building your own home. But wait, there's more. You'll be signing numerous legal documents as you build your own home. Stuff like contracts, loan papers, deeds, notes, and change orders. What are you signing—and should you be signing it? Let's take a look.

Signing Contracts

A contract is an exchange of promises. Someone promises to do something for you and you promise to pay them for doing it, for example. Once all the terms of the contract are fulfilled, it is "executed."

In most states, contracts that involve real estate must be in writing. Even though just the land is actually real estate at this point, the house will become so, once it is built. So construction typically follows real estate laws.

Must you use a lawyer to draw up a contract? No. You can write one yourself, and if you meet legal conditions, it is enforceable. Like amateur house plans, an amateur contract might not include all the important details, but it might be adequate.

What's needed to make a contract legal?

◆ There must be two or more parties.

◆ They must mutually agree.

◆ There must be some form of payment, called consideration.

That's it. So if you and a contractor (two parties) mutually agree that he will build "a house" for a price (consideration) and put it in writing, it's a legally binding contract. Of course, the contractor would be unlikely to sign such a document until the size, location, and other terms were specified in the contract. And you would need to include how and when the money will be paid and what will happen if the house isn't built as planned.

Fortunately, there are residential building contract forms available that include most of the terms that need to be mutually agreed upon. Blank spaces allow for pricing and other specifics. You can get these forms at larger office-supply retailers, your lender, and even the contractor(s). Or you can contact an attorney who will draw up a (probably better) contract—for a fee. Remember that you're dealing with many thousands of dollars here, and with your "dream." So spend a few extra bucks and get the best contract you can afford.

> **CAUTION**
>
> **Code Red** _____
>
> Consider Ramsey's Rule of Contracts: There is not enough paper in the world to write a binding contract with someone you don't trust!

Good Deeds

In Chapter 10, I talked about financing the construction of your house. That, too, will involve contracts. It also involves deeds and other public documents.

A deed is a document that conveys rights to land (and any structures on it) from one party to another. A quit-claim deed says, "I give you all the rights I have." A warranty deed says, "I give you my rights *and* warranty that I really do have those rights."

So, what's a deed of trust? It's a deed that involves a third party, a trustee, who will hold title until something specified happens. For example, you pay off the house and the trustee will give you (record) the deed. Another name for a deed of trust is a mortgage.

What about title insurance? One of the tasks of title insurance companies is to research the ownership and title for specific parcels of land, to make sure that any prior claims are known. Maybe the third owner back sold the mining rights to his brother-in-law. You should know this. And, if you buy title insurance, the company will defend claims against the title so that you don't have to.

What's a Mechanics Lien?

You're building a house, not a car. So, why should you care about a mechanics lien? Anyone who helps you build your house, investing time and/or money, is considered a mechanic by law. And a lien is a piece of paper that offers your property as security until payment for services is made.

For example, the plumber comes out and puts in all the plumbing and fixtures in your beautiful new house, but you run out of money and can't pay him. If the plumber was smart, he had you sign a mechanics lien before starting work. He can sue you for the money and force the house (with his plumbing) to be sold to pay him off.

If the lender is smart (and she is), draws may not be paid to the plumber until the bill is submitted along with a lien release. The plumber gets the check and simultaneously releases any rights to sue for the house.

A mechanics lien is a little better than a typical lien (at least for the mechanic) because it says, "Hey, I participated in the construction of this house and should get priority in payment over other types of creditors." So, who are all these mechanics? You, the lender, the general contractor (if any), subcontractors, suppliers, and—everyone's favorite—the IRS.

If you sign a mechanics lien, make sure you understand what it's saying and who gets priority. And make sure, when the work is done, that the contractor signs a lien waiver when the check is disbursed.

Code Red

Your contractor gets the check for the plumber, but doesn't pay him, so the plumber files a mechanics lien on your house. Ouch! Make sure your contracts cover this potential problem. Alternatively, have the lender or an escrow officer pay the bills and draws to make sure it's all legal.

Changing Your Mind

There's one more important legal document that you must know about. It's called a change order. In an ideal world, the house planned on paper is the one that will be built—exactly to specifications. However, as it is being built, you or the GC may want to make a change or fix an error in the plans. Someone can just go ahead and make the change, but what if it costs more money, or saves some money, or changes the structure?

A change order is a legal document that says, "I change my mind. Here's what I now want." Who cares? You should! What if Aunt Betty stops by the building site and tells a contractor, "They won't have any kids for a while, so don't build that third bedroom"? Hey, even if Aunt Betty is paying for the house, someone is going to have to write out a change order and get some signatures. If your agreement with the contractor says you're the only one who authorizes change orders, Aunt Betty is out of luck.

What's in a change order?

◆ From (you or another authorized person, such as an architect)

◆ To (the contractor or contractors)

◆ Building location, job number, and date of the change

◆ A comprehensive description of the change referring to specific plans and drawings

◆ The price of the change and how it will impact the construction schedule

◆ A statement of acceptance

◆ Signatures of the appropriate parties who are authorized (by the building contract) to make changes

Copies of change orders should go to everyone involved. That means you and any contractors who do the work, the lender, and other parties to the original building contract. If the change is significant, the local building department may also require a copy of the change order. You'll learn more about making changes in the next chapter.

Now that you've learned about codes, permits, and other concerns, Home-Building Law 101 class is dismissed. However, there *will* be a quiz!

The Least You Need to Know

◆ Building codes are written and enforced for the health and safety of building occupants.

◆ Local building departments can help you understand and comply with local building codes.

◆ Make sure you understand building and other contracts before you sign them.

◆ Get contractors to release their lien rights as you pay them.

Chapter 12

Hiring a Contractor

In This Chapter

◆ Understanding general contractors

◆ Finding and hiring the best GC for your house

◆ Developing a strong working relationship with your GC

◆ Builder beware: knowing what to look out for

Will you be your own general contractor or hire one? Before you make up your mind, you should take a look inside the world of residential contracting. What does the GC actually do? What are the job's requirements? What kind of credentials should a qualified GC have?

Those are some of the questions that I'll answer in this chapter. In addition, you'll learn what questions to ask when interviewing GCs, how to get competitive bids, and how to motivate the GC you hire.

If you're still not sure about hiring a GC after this chapter, the next one will show you how to be your own contractor. You're covered!

What Does It Take to Be a GC?

Professional folks who manage home construction use a variety of titles, including general contractor, residential contractor, construction manager, and builder. Take your pick. General contractor is one of the more popular titles because that's what most state licensing agencies call them. (My license says "General Building Contractor with Home Improvement Certification.")

What does a general contractor do for his or her money? Of course, that depends on what you hire the GC to do. What's included in the contractor agreement? In most cases, your building contractor will do the following:

◆ Supervise all aspects of the work done at the building site.

◆ Hire, supervise, pay, and fire subcontractors as needed to get the job done.

◆ Buy all materials and supplies needed in construction.

◆ Make sure that the building department inspects and approves the site.

◆ Coordinate getting building permits and any variances.

◆ Make sure that all subcontractors are legal and have the needed insurance (like worker's compensation insurance).

Typical owner/contractor agreement.

(Made E-Z Products; www.MadeE-Z.com)

OWNER/CONTRACTOR AGREEMENT

THIS AGREEMENT, made this _____ day of _____ A.D. 20 ___ by and between _____ hereinafter called the Owner, and _____ hereinafter called the Contractor.

For the consideration hereinafter named, the said Owner covenants and agrees with said Contractor, as follows:

FIRST. The Contractor agrees to furnish all material and perform all work necessary to complete the _____

for the above named structure, according to the plans and specifications (details thereof to be furnished as needed) of _____ Architect, and to the full satisfaction of said Architect or Owner.

SECOND. The Contractor agrees to promptly begin said work as soon as notified by said Architect or Owner, and to complete the work as follows: _____

THIRD. The Contractor shall take out and pay for Workmen's Compensation and Public Liability Insurance, also Property Damage and all other necessary insurance, as required by the Owner, Architect or by the State in which this work is performed.

FOURTH. The Contractor shall pay all Sales Taxes, Old Age Benefit and Unemployment Compensation Taxes upon the material and labor furnished under this contract, as required by the United States Government and the State in which this work is performed.

FIFTH. No extra work or changes under this contract will be recognized or paid for, unless agreed to in writing before the work is done or the changes made.

SIXTH. This contract shall not be assigned by the Contractor without first obtaining permission in writing from the Architect or Owner. All Sub-contracts shall be subject to the approval of the Architect or Owner.

IN CONSIDERATION WHEREOF, the said Owner agrees that he will pay to the said Contractor, $ _____ for said materials and work, said amount to be paid as follows:

The contractor and the Owner for themselves, their successors, executors, administrators and assigns, hereby agree to the full performance of the covenants of this agreement.

Witnesses:

_____   _____ Owner.

_____   _____ Contractor.

E-Z CONTRACTORS FORMS FORM NO. EZ 106

For this effort, the GC typically gets 15 to 20 percent of the total value of the project. He subtotals the costs of materials and subcontractors, and then adds a management fee. Alternatively, you could hire the GC to manage the project for a specified hourly rate. Or you might agree upon a lump sum. The advantage to paying a percentage is that there are fewer hidden costs, but it encourages the GC to spend more. An hourly rate works well for everyone if you have a way of verifying the GC's time invoice. A lump sum can keep you within a budget, but may encourage the GC to cut corners to remain within the limit. Each payment method has its own advantages and disadvantages.

So what does it take to be a licensed general contractor? It's said that someone entering the building trade must fill up a bowl with marbles, and then remove one for each month he survives the job. Once he has lost all of his marbles, he's qualified to be a contractor!

Actually, contractors are licensed based on knowledge, experience, and other assets. For example, in California, a licensed general contractor must …

◆ Be 18 years of age or older.

◆ Prove at least four full years of experience as a journeyman, foreman, supervisor, or contractor in the appropriate classification.

- Pass a written test.
- Have a specific amount of operating capital (currently more than $2,500).
- Register a contractor's bond or cash deposit ($7,500).
- Purchase a worker's compensation insurance package (expensive in the building trades).
- Pay the examination fee ($250) and licensing fee ($150) plus bi-annual renewal fee ($300).

California, as in many other states, has more than one type of general contractor. There's a general engineering contractor that requires specialized engineering knowledge and skills. And there's a general building contractor for constructing houses and other shelters. There are also numerous specialty contractors for specific trades (such as lathing contractor or carpentry contractor). The requirements for these folks will be covered in Chapter 14.

The four years' experience can include education, such as college or trade-school courses. But it must include at least one year of full-time experience managing a construction project.

If you do decide to go for your general contractor's license someday, your state licensing board can tell you what books to read and what classes to take. The licensing examination will probably have two sections. The first is on the specific trade, such as shelter construction. The second is on law and business topics like project management, bookkeeping, bidding, safety, contracts, liens, insurance, and similar topics. Questions are typically true or false and they offer numerous test-coaching services to help people prepare for and pass the state contractor's license examination.

Picking Your Partner

Now you know approximately what a general contractor does for his or her fee—manages the building project. And you know what it takes to be a general contractor: knowledge, experience, and some assets. So how are you going to find the *best* contractor you can afford to build your home? Fortunately, with more than one million contractors licensed in the United States, you probably won't have to settle for a GC who meets only the minimum requirements.

The search begins early. In fact, even before you've drawn up house plans, you're already looking for a qualified general contractor to manage the project. You don't hire a specific contractor until you've interviewed many, but you've started looking.

On the Level

In most states, owner-builders are exempt from licensing requirements when they build or improve on their own property, if they do the work themselves or hire employees to do so—*if* the structure is not intended or offered for sale for one year after completion. (More on this topic in Chapter 13.)

Referrals are the best place to find GCs. Ask people you know who have had homes built in the area recently. Talk with local building material suppliers. Ask your lender and any subcontractors you know for recommendations. What you're looking for at this point is a comprehensive list of builder candidates. Start writing them into your Home Book. You can weed them out later by calling for qualifications and availability, and then following up with interviews of the best candidates.

Finally, you've narrowed the field down to a handful that you ask to bid on your project. Availability is a key factor in many areas. You may need to build during the summer months, but learn that the best contractors are too busy on their own projects to take on anything new until the winter or next spring. In fact, if a candidate isn't busy when others are, maybe there's something you don't know. Or it could just be that the GC is pickier about what he or she takes on.

How do you interview a contractor? Face to face, if possible. At lunch, during a long coffee break, or at an office or job-site appointment. Of those, seeing a contractor at a job site can be enlightening, as long as your presence isn't distracting.

On the Level

In the movie *The Naked Gun* starring Leslie Nielsen, there's a scene in which Vincent Ludwig (Ricardo Montalban) attempts to kill the Queen of England during a baseball game. Jane (Priscilla Presley) asks, "How can you *do* this?" Ludwig answers, "It is easy, my dear. I spent two years as a building contractor!"

Here are some questions you should ask candidates:

- What can you tell me about your license and construction experience?
- What type of contracting license do you have and what is the license number?
- What experience do you have with owner-builders?
- What references can you give me?
- Can you post a performance bond (an insurance policy guaranteeing that you'll finish the job)? If not, why?
- May I see your worker's comp policy?
- How do you go about hiring subcontractors?

- How do you schedule a house-building project?
- Do you typically meet your building schedules?
- What lenders do you prefer to work with? Why?
- What suppliers do you prefer to work with? Why?
- What things do you need to make a firm bid on this project?
- How do you develop an accurate construction bid?
- Can I have my attorney look over your standard building contract?
- If I hire you, what's the best way to discuss the project with you?

Of course, you could ask many more questions, depending on your house plans and how you will participate in the building process. Remember, you want to ask open-ended questions that aren't answered by a simple "yes" or "no." You want to hear explanations and find out how well the person communicates.

Write down all the questions in your Home Book so that you can ask each candidate the same questions and make sure you keep track of the answers. The question you forget to ask may be the one you later wished you had asked!

The contractor's license number will come in handy. License contractors typically keep a file that you can access if you know the contractor's license number. You can do without it, but it makes the job much easier if you know the number. The file will have useful information about when and where the license was issued, as well as show any complaints or commendations filed. Keep in mind that a board of retired contractors typically hears the complaints against contractors!

Calling for Bids

You've found and interviewed lots of contractor candidates, and have selected three to six you think you can work well with. Time to talk money! The bidding process is critical to you and to the contractors. You don't want them to make too much money on the project, nor do you want them to lose so much that they have to cut corners or not finish the job. Ask for bid forms from each contractor or, if possible, standardize on a single bid form to make your review process easier.

Code Red

Don't assume that the license number the contractor gives you is accurate. It may be an inactive license or issued to someone else. Or numbers may have been transposed. Check it out before investing thousands of dollars with this person.

The key to accurate bids is accurate building specifications. That is, contractors don't have to guess (to their advantage) when working with clear specs. They know exactly how much the materials will cost, how many hours it will take for each stage, and how long until it is done. So, your first job is to make sure that each general contractor has a complete set of construction plans (see Chapter 11). The cost of extra plans is small compared to the total cost of the project, so if you really want an accurate bid from a contractor, make sure that he or she has all the plan prints needed.

Of course, the contractor may use rules of thumb anyway. "Let's see, 2,100 square feet at $125 a square foot is $262,500. Sounds about right." Fortunately, you asked each contractor about how he or she approached bidding, and weeded out any who use such coarse rules. Those kinds of approaches always favor the contractor.

Be sure you know what's included in the bids. Building permit fees? Utility connection fees? Travel expenses? Site security? Cleanup charges? Also ask for a start and finish date in the schedule. It will make a difference as you pay interest on the construction loan or need to schedule the rest of your life around this project.

Get job responsibilities in writing on the bid. Who will call for inspections? Who notifies the lender to release a draw? How will changes be handled? The building contract must answer these questions, so ask the contractor to include them in the bid.

Apples to apples. That's what you're looking to compare. When all the bids arrive, you want to be able to look at them side by side and see which is higher or lower in specific construction tasks.

Reviewed all the bids and figured out which one is best? You're not done yet. Now it's time to negotiate. Maybe your favorite GC should use another bidder's subcontractor for flooring for a lower price. Or you will get a better price if contractor B resubmits an hourly rate bid rather than costs-plus-fee bid.

You can narrow the contractors to two or three of the best bids, and ask for a rebid based on minor changes. You can also tell contractors what you're looking for in the final bid; for example, "How can I get a final bid that's 10 percent less without cutting quality?" Or "I need a rebid based on my doing all the framing." Or "I want your quality at his price."

Be careful. Don't get cheap. If you've developed a working relationship with these bidders, you have an idea what it will take to get a fair bid. And that's what you want.

Getting Your GC to Work Hard for You

Found the right contractor to build your house? Great! Let's get it in writing. A typical builder agreement will include some important specifics. Attached to the agreement (by reference) will be the building plans and probably the final bid. The agreement itself will include the following:

♦ Contract date and parties

♦ Starting and completion dates

♦ Conditions (who does what, who pays for what, who approves what, how changes are made and paid for, and so on)

♦ Contract sum

♦ Progress payments (matching the lender's draw schedule, if any)

♦ Terms of final payment

♦ Contract termination terms

♦ Anything else agreed upon that should be in writing

Don't sign anything under pressure. If you're not comfortable with the price, terms, and conditions, stop and think it through. Will this contract take you to your goal? What's the worst that can happen? Is it covered by the contract? If it will make you more comfortable (and it probably will), have an attorney review the building agreement prior to and during final signing.

Your Relationship with the GC

You've signed a building agreement with a professional general contractor whom you trust to do the job on schedule and on budget. Now you must maintain that relationship of trust, remembering that it is based on mutual benefits. You want a house; he wants the money. You don't want legal or financial troubles, and neither does he. You want a contractor you can recommend and he wants a reference. How can you maintain and benefit from this relationship?

Ka-ching!

What motivates the typical contractor? Making money! What motivates the typical owner? Saving money! To encourage a contractor to build for less than the bid, offer to split the savings with him—as long as the materials and workmanship meet specifications.

Communication is the key. That doesn't mean you have to talk on the phone or at the site all day. But it does mean you need to know what's going on to your comfort level. You may agree, for instance, that if you don't hear from the contractor between five and seven each evening, then the day went as scheduled and previously agreed. If there is a problem that you can help solve, the contractor will let you know during that call. "Can you pick up the building permit tomorrow?"

If you're not home much, make sure you have a telephone answering system or service where the contractor and others can leave detailed messages. In fact, you may not speak "live" to each other for weeks.

You'll get additional ideas on how to keep communications going as you read Part 4, on the full construction process. For now, start developing a working relationship with your general contractor. It will be like a marriage—with money replacing intimacy.

The most crucial part of the working relationship is establishing how changes are approved and made. Changes will happen. The specific window may be backordered. The foundation plans may not be correct, and now you must change the framing plans. You may decide to upgrade or downgrade the roofing materials or add a small dormer.

Who approves changes? Who designs the change? Who pays for it? These are important questions. Changes occur during three periods. When the changes are made impacts who makes them and how much they cost:

- During planning, when simple design changes cost some research and some redrawing
- During installation, when some time and materials are lost
- After installation, when more time and more materials are lost

Of course, it's preferable to make all changes during design. It's less expensive. Design changes can even save you some money. For example, you may decide that the staircase is too complex and that a simpler design will be as functional and ornamental at a much lower cost. Changing the staircase during construction will be more expensive. And changing it after the stair system is in can be quite costly. It can even delay other construction steps.

The moral is for you and others to review the plans many times, looking for potential changes before finalizing the plans. The other moral is to keep good communications going with the contractor so that you can efficiently participate in needed changes.

And remember that any change you make to a home design impacts other parts of the house. Changing ceiling height also changes stairs and door headers. Adding a masonry fireplace late means cutting through the subfloor to install a foundation for it. Moving a wall means making sure the roof load is properly supported. It can get complicated!

Contract Change Order

FROM:

NO.

DATE

JOB

TO:

CONTRACT JOB NO.

PREPARED BY

WORK TO
BEGIN BY / /

WORK TO BE
COMPLETED BY / /

The work covered by this order shall be performed under the same Terms and conditions as that included in the Original Contract.

Changes Approved

BY

BY

®E-Z CONTRACTORS FORMS FORM NO. EZ 114

By: _____

PREVIOUS CONTRACT AMOUNT	$	0
AMOUNT OF THIS ORDER	$	0
TOTAL CONTRACT AND EXTRAS	$	

Typical contract change order form.

(Made E-Z Products; www.MadeE-Z.com)

Builder Beware

Someday we may all live in an ideal world. Meantime, there are folks out there who will take advantage of other people's lack of specialized knowledge. So what can go wrong in hiring and dealing with building contractors?

◆ Poorly written or verbal contracts allow contractors to walk away with other people's money.

◆ Verbal promises are not kept.

◆ Contractors substitute inferior materials and pocket the difference.

◆ Contractors receive hiring fees from winning subcontractors.

◆ Unscrupulous inspectors are bribed to sign inaccurate inspections.

◆ Unlicensed and unbonded contractors and subs are used, increasing owner liability.

◆ Unauthorized changes are made to plans during construction.

◆ Liens are not removed after payment, increasing owner liability.

◆ Workers not covered by worker's compensation insurance are injured at your building site.

◆ Available discounts on materials are not passed on to the owner as agreed.

You get the picture. Fortunately, learning how homes are built and how to hire and work with reputable contractors can save you both money and frustration. You can save even more money by being your own contractor. I'll tell you how in the next chapter.

The Least You Need to Know

- ◆ General contractors are state licensed after passing minimum requirements and a test.
- ◆ Finding and hiring a qualified GC means building a relationship of trust.
- ◆ You can tell a lot about how a contractor works as you review his or her construction bids.
- ◆ Beware of hiring unqualified contractors who can increase building problems and your liability.

Being Your Own Contractor

In This Chapter

◆ You *can* be your own contractor

◆ What owner-contractors do

◆ Dealing with building inspectors

◆ Making sure you and others are insured

"Me, be my own c-c-c-c-contractor?" This is where some folks get nervous. Especially those who have figured out what a contractor does. It's work. Lots of work. Fortunately, you have help. The tools you need to be your own contractor are in this book. So the point of this chapter is to help you relax—with facts. You'll learn more of what an owner-builder does and what you need to do to build your own home.

The payoff is two-fold. First, you save money. Depending on how much you do yourself, you can save many thousands of dollars by being your own contractor. Second, you earn bragging rights. You can say, "I built this home." And even if you had help, you'll be right!

Can You Really Be Your Own GC?

In the previous chapter, you learned that a general contractor license requires real-life construction and supervision experience. A GC must have credentials before hiring out to others. But, with some exceptions, you don't need a general contractor license to build your own home. Why? Because it's *your* home. And this is the land of the relatively free. If you're willing and able to abide by locally accepted building codes and construction rules, you don't need a license to build a residence you will live in—at least for a while.

Having said that, remember that anyone who finances the construction of your house will want to make sure you're qualified to manage construction. But the lender doesn't require you to have a license, either. If your lender requires that a licensed general contractor manage the job, and you want to build it yourself, find another lender. Many of them have experience with owner-builders who can illustrate that they are good investments.

Understanding the Contractor's Job

So let me remind you of what a general or building contractor does: He or she manages. A GC doesn't even have to pound nails or pull wires for his or her money. The GC is in charge. A general contractor …

- ◆ Finds reputable material suppliers.
- ◆ Negotiates discounts.
- ◆ Accepts material deliveries at the building site.
- ◆ Finds qualified subcontractors.
- ◆ Negotiates contracts.

On the Level

Want to be a general contractor some day? In most states, your experience as an owner-builder counts toward requirements for an official contractor's license!

- ◆ Inspects subcontractors' work.
- ◆ Calls for building inspections and meets inspectors.
- ◆ Participates in the building process as qualified and desired.
- ◆ Supervises subcontractors on the job, answering questions about plans.
- ◆ Authorizes payments for materials and subcontractors as needed.
- ◆ Provides worker's compensation and job insurance as needed.
- ◆ Leans against a pickup truck while talking on a cell phone.

Does this fit your skills and lifestyle? Can you afford to take the time to be on the job site nearly every day? Can you take a hiatus from your job to build your house? Can your partner do so?

If you're wondering how you, an owner-builder, can find qualified subs, remember: *Good people know good people.* That is, once you've found a qualified and reputable subcontractor (who may serve as your construction advisor), ask him or her to recommend other subs. Chapter 14 offers dozens of suggestions on how to hire, manage, and motivate subcontractors.

Considering All Your Options

Thousands of houses have been built by owners who are only on the site during the weekend, supervising material deliveries and subs, who have weekday jobs just like the owner does. You have lots of options if you want to be your own contractor on your home. Here are a few:

- ◆ Take a leave of absence from your job during the most crucial part of construction.
- ◆ Hire a trusted subcontractor to be at the site when you can't.

Ka-ching!

Hiring a co-contractor? Because much of modern residential construction involves framing (foundation, walls, roof), the framing contractor will be at the building site more than any other. Consider hiring this person as your construction advisor and to be on-site when you can't.

- ◆ Share construction responsibilities with your husband, wife, partner, significant other, adult child, retired parent, or trusted relative.
- ◆ Schedule work and deliveries at the site for weekends only.
- ◆ Rearrange your day job to have two or more weekdays off, and work the weekends.
- ◆ Accept a current or new job with an evening or midnight shift so that you can be at the site during the day.
- ◆ Hire a licensed contractor to advise you on construction management, and be at the site when you cannot.
- ◆ Securely install a cell phone at the building site, and let everyone know your work telephone number so that they can call you as needed.

One word of caution: If you're financing the construction of your house, whoever is loaning you the money will want assurances that you know what the heck you're doing. If you do, write down your qualifications and experience for the lender. If you don't have much experience, hire someone who does to advise you and to set the lender's mind at ease.

Do You *Really* Want to Do This?

You've probably figured out by now that you *can* build your own home as your own general contractor. So the next question is, do you *want* to? You're obviously going to save some money. By being your own contractor you can save as much as 25 percent on the cost of your house—though most owner-builders spend the same and get more.

What will you be giving up? Time. Where will you get the time? Maybe by taking time off from another job or responsibility. Or from a relationship. Actually, building your own home can give you time over the long run. Your new house may mean you spend more time at home or share the building process with family and friends or add marketable skills to your resumé. So you're not really giving up time, because you're trading it for something else. As you think about being your own contractor, be sure that you get a good trade.

If your answer to the question "Do you want to?" is still "yes," congratulations! Building your own home is satisfying business.

That's right. It's a "business"—and you must treat it like a business to ensure its success. That is, take the job seriously, plan on investing lots of time and money into it, make it a high priority in your life, and expect a profit from it. You may reinvest that profit right back into the house, but it is yours. You earn it!

Reality check! You've decided that, although you want to save some money, you're just not cut out to be a contractor, nor do you want to manage your subs. You'd just as soon hire it done. For some folks, that's the smartest move. Why take off time from a well-paying job to save less than you would earn? Or maybe you have personal or family health issues that will keep you from getting the job done well.

Isn't there *anything* you can still do to reduce costs and increase pride of ownership? Certainly! There are many things owners decide to do in the construction of their home that improve the quality and/or lower the cost. What they do depends on their skills, time, and health. For example, owners can take on any of these jobs:

- Painting and wallpapering
- Helping others install doors, windows, flooring, cabinets, trim, and similar materials
- Installing lighting fixtures (with directions from the package)
- Cleaning up the building site, removing scrap lumber, nails, drywall trim, and so on
- Providing site security against theft and vandalism
- Picking up materials as needed by contractors
- Keeping track of all draws, expenses, bids, deliveries, and inspections for the contractor
- Being a go-fer as needed by any of the contractors or laborers at the job site

Remember, if your contractors and craftspeople are being paid by the hour, anything you do that saves them time also saves you money!

On the Level

If you think you want to be your own contractor but don't feel you have the skills, consider going to school! Contact your local community college for construction trade classes, or enroll in one of the owner-builder schools (see Appendix B) located around the country.

Working With Inspectors

Inspectors have the power to stop the construction of your new home and send everyone away at the cost of thousands of dollars. Or at least that's the fear. Actually, inspectors are your allies in getting your home built. The inspector's job is to make sure the house is safe and healthy for occupancy. He or she periodically visits the construction site to make sure that the residence is being built according to the plan approved by the building department. And since you, too, want it built according to those plans, the inspector is your friend. Or at least not the enemy.

On the Level

Many building departments prefer to hire college grads who have taken extensive coursework in building codes rather than hire from the building trades. They want to train them their own way.

As you get a building permit, ask about inspections and, if possible, meet one or more of the inspectors. Find out how much advance time you need to call for an inspection and what work, if any, you can continue doing prior to the inspection. Once you schedule an inspection of a specific component, don't do any more work on that component until the inspection is done. That is, hold off installing roofing until after the inspector has inspected the sheathing installation and nailing. Also find out what happens if the inspector doesn't sign off on the inspection. What must you do before calling for a reinspection? Who do you call? What if you dispute the inspection?

As the owner-builder, you will also deal with the lender's inspector(s). In some cases, the lender may accept the county or city building inspector's certification. Other lenders may have their own inspector visit the site. In many cases, draw schedules don't coincide with building inspections.

Why does the lender want the house inspected? Because the lender pays contractors and material suppliers from draws based on what's been done. Chapter 10 describes the draw process for a typical residential construction loan.

As the lender's inspector sees that specific milestones have been reached, she or he authorizes payment of an amount specified in the draw schedule. The lender may cut the check or have an escrow officer or escrow company do it. In any case, as the owner-builder, the money may be deposited in your building account from which you can pay contractors and suppliers. So it's important that you keep track of expenses as well as income from draws and other sources, and manage the cash flow. You don't want to face the day when you have a $50,000 draw and $100,000 in bills.

As your own contractor, learn how to keep good records and manage a checkbook. If you have a computer, invest in one of the checking account software programs (such as Quicken, QuickBooks, or Money) and learn how to use it. In fact, your lender may require that you do so.

The toughest inspector at the job site might be you. You're the owner. You want to make sure that everything is done not only to code and plans, but also to your image of the finished house. Going forward without a thorough inspection can cost lots of time and money. Here are some of the most important points to look for in the construction process:

Code Red

Incorrectly installed flashing in one new house cost the contractor 26 new sheets of drywall—and lots of time and money—as rainwater cascaded through a leaky roof and down walls!

- ◆ **Elevation.** Make sure that the site has been surveyed, excavated, and graded so that the house foundation will be at the proper location and height.
- ◆ **Foundation.** Verify that the footings and foundation forms are accurately set and of the correct size to the plan.
- ◆ **Rough-ins.** Make sure that any required plumbing or wiring to be installed within the concrete slab or foundation walls is installed in the foundation.
- ◆ **Doors and windows.** Check that the appropriate sizes of openings are built into the walls for doors and windows.

- ◆ **Utility runs.** Make sure that all plumbing pipes, electrical wiring, HVAC, and other utilities are installed in the walls, ceiling, and floors correctly before closing up the walls.
- ◆ **Roof flashing.** Check that the roof seal is properly installed for run-off.

Getting Adequate Insurance Coverage

Insurance is a sleep aid. It helps you minimize worry about financial loss from a potential hazard, and thus you sleep better. The price of insurance is the premium. Reading the policy also can be a sleep aid.

Builder's Risk Insurance

Once you've moved into your house, you'll want to get a homeowner's policy to cover potential loss from hail, tornadoes, or other natural acts blamed on God. Or for things that humans do, such as burn things, steal things, or injure themselves on your property.

But what about an insurance policy for potential losses while you're building your home (or having it built)? It's called a builder's risk policy, and you can probably purchase it through your regular insurance agent. It covers losses from acts of humans or nature during construction including fire, hurricane, and theft. It *doesn't* cover the loss if an unscrupulous contractor takes off with your money or materials. Once the house is finished and an occupancy permit is issued, the builder's risk policy will probably convert to a homeowner's policy, and will then cover additional things like all of your furniture and belongings.

Your lender may require a builder's risk policy for financing. In fact, some will automatically build one into the loan package. Be careful, though, as it will probably only pay for the lender's loss, not yours. Make sure you know what it will and won't cover. If needed, add riders or additional terms to the policy to insure anything you might lose. Also, consider whether you need a rider to cover theft or vandalism at the building site.

General Liability Insurance

General liability (GL) insurance covers other things that can happen during construction. Which things? That depends on what insurance you purchase. For example, the painting contractor sprays the exterior on a windy day and droplets dust a neighbor's car. (True story!) Who covers the cost of detailing or repainting the car? The painting contractor? Your builder's risk policy? You? A general liability insurance policy? (In this case, the contractor's general liability insurance.)

The time to act is now, before anything happens. If you're using a general contractor, make sure he or she has general liability insurance, and find out what it covers. If you are your own contractor, get GL insurance before starting construction.

Worker's Compensation Insurance

What about a carpenter who breaks his thumb with a hammer? Or a roofer who slips and falls off the roof? Construction is dangerous work. Worker's compensation insurance covers workers while on the job. So worker's comp would cover the carpenter or roofer in the previous examples, and the state would pay for it. That's because the employer (the contractor or you) is paying worker's compensation insurance premiums to the state as required by law.

Code Red _____

As you can imagine, worker's compensation insurance premiums in the hazardous construction trade aren't cheap. It's one place where unscrupulous contractors shave costs—they just don't pay them. So make sure that any contractors you use give you a worker's comp certificate issued by their insurance company. A claim against the contractor, even if it isn't directed at you, may tie up your home.

The alternative is for the injured worker to sue the contractor and/or you! By paying worker's comp insurance premiums, you are insuring yourself against a lawsuit for these injuries. So you want to make sure that any contractors who work for you have and are paying for worker's comp insurance. How do you know? You can require evidence, in the form of certificates, that the contractor is covered—and so you are not liable. You don't really want an injured worker to own your new house, do you?

The Least You Need to Know

- You can successfully be your own contractor if you can manage the multiple projects needed to build your own home.
- Before accepting the task of building your own home, make sure you understand the physical, mental, emotional, and financial costs.
- Building inspectors can make your job easier or more difficult, somewhat depending on how responsive and cooperative you are.
- Make sure that anyone hired to help you build your home is adequately insured.

Chapter 14

Hiring Subcontractors and Laborers

In This Chapter

♦ Learning what the various subcontractors do—and how to find qualified people

♦ The importance of being licensed

♦ Selecting the best and most cost-effective contractors

♦ Motivating and managing contractors

♦ Hiring helpful laborers

Getting excited about building your own home? Great! It's an exhilarating adventure that will be the world's best job one day and the worst job the next. Or maybe both on the same day.

If you plan to be your own contractor, a big part of your daily job will be to hire, manage, motivate, and, if needed, fire subcontractors and laborers. That's the topic of this chapter.

Hiring the Best and the Brightest

It's a crazy world out there. Lots of folks will tell you, "Yeah, I can do that job—for less!" Then, once done, you discover exactly *why* they charge less than anyone else. They *do* less! So how can you, without the advantage of friends in the construction business, find some friends in the construction business?

What contractors will you need to help you build your home? That depends on the design of your house and what you plan to do yourself. Let's take a quick look at what contractors (only called *sub*contractors when working for another contractor) do.

Grading Contractor

Grading contractors make the earth move! They add or remove soil to prepare the building site for the foundation. That may include excavation, filling, and contouring or grading the site. It may also include cutting in a driveway.

The first step is for the grading contractor or a surveyor to "shoot the grade," or measure where the boundaries, structures, and other points are. You don't want your house to be located in the wrong spot. Then, using heavy equipment, the grading contractor prepares the site for the foundation walls and/or slab, any septic systems, water lines, driveways, sidewalks, and other stuff in your plans.

Chapters 17 and 18 offer more complete info on what the grading contractor does.

> **Ka-ching!**
>
> To find good specialty contractors in your area, ask the local building department for the locations of residential homes currently under construction in your area. Then visit the sites during times when the crews will be there and watch construction, especially around quitting time. If possible, speak to some of the workers or contractors, letting them know you are building your own home and looking for qualified contractors. Take it from there as appropriate. Also, contact the state contractor licensing board to check up on your contractors before you hire them. You can find licensing boards in the state offices section of local phone books.

Foundation Contractor

The foundation contractor is responsible for preparing the site for and installation of the foundation by the approved building plans. The foundation contractor may work with the grading contractor to dig for footings, or may have his own crew do it.

The foundation contractor will install forms, install reinforcement bar (rebar), work with plumbers or other contractors who need to bury their utilities in the concrete, and get everything for the concrete. The foundation contractor may hire a concrete contractor to pour and finish the concrete, or he may have his crew do it.

Chapter 18 will get more specific about installing the foundation.

Plumbing Contractor

The plumber might rough-in the plumbing system or start early in the construction process. The plumbing contractor will install runs or horizontal pipes and risers or vertical pipes for the water, sewer, and drain systems.

If something will later be built over the top of these pipes (such as concrete slabs), the plumbing contractor will do so as the foundation is being built. Otherwise, the plumbing contractor can rough-in plumbing later.

Once the framing is done, the plumbing contractor will be back to install supply, drain, and vent pipes throughout the house. Finally, the plumbing contractor will return after the walls are closed to install plumbing fixtures and hardware.

Chapter 22 covers services such as plumbing in greater detail.

> **On the Level**
>
> When you're searching for qualified electrical contractors, look for their membership in the National Electrical Contractors Association or NECA (www. necanet.org). NECA also offers referral and arbitration services.

Electrical Contractor

The electrical contractor starts at the power drop, the wires that come in from the power pole or underground tap, installing an electrical meter followed by a breaker box. From the box, wires are run throughout the house through framed walls, floors, and ceilings. Finally, the electrical contractor returns when the walls are closed in order to install outlets, switches, and cover plates.

Chapter 22 gets into more detail about installing your home's electrical system.

Framing Contractor

One of the busiest folks you'll hire, the framing contractor is responsible for erecting the structural frame of your house. This includes joists, floor decking, walls, rafters or roof trusses, and roof sheathing. The framing contractor will probably also install exterior sheathing or siding, doors, and windows.

Chapters 19 and 20 cover the flooring and framing of your house; Chapter 21 fully describes the exterior jobs.

Roofing Contractor

The roofing contractor, as you would guess, is in charge of installing the roofing materials. It includes the roofing materials (asphalt, tile, shingles, and so on) as well as flashing (edges), and boots (vent flashing). Your roofing contractor might also install rain gutters and downspouts, or you might call in another special contractor for the job.

Roofing is fully covered (pun intended) in Chapter 21.

Drywall Contractor

Drywall is a sheet material made of gypsum plaster covered in a special paper, cut into sheets of 4' × 8' or 4' × 12'. The drywall contractor nails or screws the sheets to wall framing, and then tapes the seams so that the walls look seamless. Some drywall then has a texture applied to the walls and/or ceiling. Alternatively, they plaster walls with a wet plaster installed over wood lathe or other surface. The plaster is hand-contoured, and then dries.

Installing wall covering including drywall is described in Chapter 23.

Flooring Contractor

The flooring contractor will install carpet, tile, solid or laminated wood, or other flooring materials per the building plans. Some flooring contractors specialize in one or two types of flooring, such as carpeting, and leave the others to someone else. There are specialists among specialists.

The flooring process is more fully described in Chapter 23.

Painting Contractor

Guess who paints your house? The painting contractor. (Too easy!) Even painting contractors have specialists. Some do exteriors or interiors only. Others specialize in painting high places. Still others work with special finishes or painting techniques.

Chapters 21 and 23 tell more about the painting process.

On the Level

Larger painting contractors are often members of Painting and Decorating Contractors of America (PDCA).

Mechanical Contractor

One or a group of contractors may tackle the heating, ventilation, and air conditioning (HVAC) systems. They design the systems, install ducts, install furnaces and fans, vents, and compressors. Finally, they install filters, registers, and grills.

Chapter 22 describes HVAC system installation more completely.

Other Contractors

Depending on your home's design, you may need masonry, ironwork, millwork, finish, cleanup, and other contractors. In smaller communities and rural areas, you may find that contractors take on more jobs. For example, the drywall or framing contractor may also handle the insulation. Or the framing contractor may tackle the roofing. Or the electrical contractor can bid on the HVAC system.

Making Sure Your Contractor Is Licensed

Let's discuss licenses. In Chapter 12, I told you how general contractors get their licenses in many states. Subcontractors follow a similar licensing process. For example, a contractor needs to prove to the state that he or she has the required construction trade experience. Then the candidate must pass a test, usually multiple choice, and pay a fee. He or she receives a specialty contractor's certificate or license.

Scads of specialty contractor licenses are available in the building trade. To give you an idea, they include insulation and acoustic contractor; boiler, hot water heating, and steam fitting contractor; carpentry contractor, cabinet and mill work contractor, low voltage systems contractor, concrete contractor, and, well, more than 35 others.

Finding Qualified Contractors

The biggest headache that owner-builders face is dealing with subs. Some are late, don't show up, don't bring materials or even tools, don't follow the approved building plan, or skimp on workmanship. If you can find, hire, and manage good contractors, you'll definitely earn your savings! Where can you find qualified contractors? Here are some ideas to get you started:

Code Red

Don't get burned by a bad contractor. Contact your state contractor licensing board (see Appendix B) for information in their files about the contractor. You may find a gem—or a lump of coal. And remember, the more knowledgeable an owner-builder is, the more honest the contractor will need to be.

- Referrals from trusted contractors
- Referrals from other owner-builders
- Referrals from material suppliers
- Building sites
- Telephone book business listings
- Newspaper classified service ads

The first question you'll ask any sub you interview is, "What's your state license number and specialty?" The second question should be, "What references can you provide?" Of course, no one is going to give you references who will bad-mouth them. But a contractor may give you bogus ones and hope you don't follow up. You're smart, so you'll call each one of the references before you trust the contractor with your house—and your money.

Finally, you ask the contractor to make a formal bid on your construction project. Chapter 12 offers specific steps to getting a qualified bid from contractors and subs. To summarize:

- Provide specifications and plans to each contractor you've selected to bid.
- Verify who pays for materials when and how, and verify the quality.
- Keep a log of who you've asked to bid, and follow up as needed to get the bids in.
- Make sure everyone knows your deadline for accepting preliminary bids.
- Once all bids are in, review them, and ask any that aren't clear or competitive to resubmit.
- Make sure that all bids are in writing.

Hiring good contractors can make your home construction easier. Hiring the wrong ones can make it h-e-double-hockey-sticks.

Why is hiring good contractors so important? Because, as an owner-builder, most of your job will be managing them. The rest of your job will be dealing with materials issues and getting inspections done and signed off. Additionally, you'll have to deal with prevention of theft and vandalism at the building site. Good contractors can make each of these tasks easier.

> **On the Level** _____
>
> Communication is a key to managing contractors. Make sure the contractors you interview have a land telephone, mobile or cellular telephones, a fax machine, an answering system, and other ways to reach them as needed.

Motivating Your Contractors to Do Their Best

Contractors don't typically work for the glory. We're in it for the money! Money is our primary motivator. Yes, many of us are also craftspeople who rightfully take pride in their work. Forget to compliment them and no harm is done, but forget to pay them and you'll have problems.

How are contractors typically paid? You learned about construction draws and other financial stuff in Chapter 10. The lender puts money in your account as you meet specific construction milestones. How you (or your GC) pay your contractors is between you and the contractors. Typically, 50 percent of the agreed-upon fee is paid when the initial work or rough-in is done and the balance once it is completed and passes inspection. Some GCs and owner-builders prefer to pay 45 percent on rough-in, 45 percent on completion, and the final 10 percent two to four weeks later, just to make sure there are no hidden problems. Others pay 40 percent on rough-in and 60 percent when finished. A framing or other contractor may have numerous tasks, and therefore may require more than two or three payments. If you are paying contractors by the hour, have them submit time sheets at specific milestones, such as rough-in or completion.

Another way to motivate contractors is to offer bonuses. For example, if the busy plumbing contractor will only be available for four days next week and framing must be done by then, consider offering the framing contractor a bonus to get the required job done on time. How much? It depends on how much time or money you might lose if the plumber is delayed.

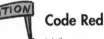

Code Red _____

Whatever you do, don't pay ahead for work not yet done! Don't pay down payments or initial fees. It removes motivation for most contractors to continue or complete a job.

Want to *really* motivate a crew to move faster? Offer cash bonuses upon completion of the work!

Managing Contractors Day to Day

The key to managing contractors is to make sure that they show up on the job as scheduled. Good luck! You will find contractors who keep to their schedules. You'll also find those who don't. Mostly, you'll find contractors who would like to stay on schedule, but things get in the way—like other jobs.

You can't prevent unforeseen problems at other construction sites (or bad weather) from changing your building schedule. However, you can motivate contractors and crews to do whatever they can to stay on schedule.

It is important to understand perspectives. It's your house. For the next three months to a year, its construction will be a very big part of your life. However, for your contractors and laborers, it's just another job. In a few months your house will be replaced by another house, and then another. Don't expect others to have the same interest in getting the job done right as you do. Do learn what motivates them and use it to your advantage.

Having problems with a specific contractor or laborer? If it's one of the contractor's employees, speak with the contractor about the problem. If it's the contractor himself, the same rule applies. Also make notes of any problems or issues, including names and times, in your Home Book.

If someone is just not working out, you do have recourses. First, make sure your written agreement with contractors covers handling problems. Can the lender or another contractor arbitrate? Is there a local contractor's association that can help?

Of course, the best way to avoid problems with contractors is to hire ones who don't make problems. That means doing your homework *before* you hire a contractor—and making sure any agreement you sign includes a process for arbitrating disputes without stopping construction.

A written agreement with each contractor you hire should include the following information:

◆ Contractor's name, address, and telephone number(s)
◆ Contractor's license number and type
◆ Contractor's tax identification number or Social Security number
◆ Verification of worker's compensation insurance (including policy number and expiration date)
◆ Job address and description
◆ Job specifications (attach plans as needed), including who orders and pays for materials
◆ Price, terms, and payment schedule
◆ Terms of arbitrating disputes
◆ Signatures and date

Hiring Good Laborers

Depending on your own skill level, you may decide to take on more of the actual construction yourself and hire laborers to help. Many owner-builders take this route. They hire people with specialized skills such as plumbers and electricians, but do the framing and painting themselves, hiring laborers to assist as needed.

Here's where you can find such laborers:

◆ Newspaper classified service ads
◆ Local employment offices
◆ Employment services
◆ Qualified friends and relatives
◆ Moonlighting construction workers
◆ College employment offices (especially those with construction trade classes)
◆ Referrals from material suppliers

First, figure out what you need and write a description. For example, "Need experienced nailer to help frame an owner-built home. Must be able to read plans and specs." Or "Need strong young person to help with foundation and cement work for two weeks." A clear description will help direct you to the best place to find such a person.

How much you pay your laborer depends, of course, on the value of the work done and the local labor market. Ask your construction advisor, contractors, material suppliers, employment offices, and labor halls what it will take to get qualified help. Then plan to pay 10 to 20 percent more for good help.

Make sure you read the section on worker's compensation insurance in Chapter 13!

The Least You Need to Know

- ◆ Numerous qualified contractors and tradespeople are available who can help you build your home without robbing your bank.

- ◆ The more you know about the building process, the easier it is to find and hire qualified contractors to do the work.

- ◆ You want to save money; contractors want to make money. Find a compromise with qualified contractors.

- ◆ Make sure any contractors you hire are licensed.

- ◆ One of the most difficult tasks of building is managing contractors and getting them to do the job you hired them for. Learn what motivates them and use it to your advantage!

Two-Off-Ten: Hiring Suppliers

In This Chapter

- ◆ Gathering your suppliers
- ◆ How to hire a good architect and engineer
- ◆ Finding a reputable materials supplier
- ◆ Keeping your suppliers working hard for you

As a senator once said, "A million here. A million there. After a while we're talking about some *real* money!" You'll be spending *real* money—many thousands of dollars—as you build your own home. You'll be paying an architect or plan service, hiring an engineer, ordering materials, and more.

This chapter is about getting the best advice and materials at the lowest prices delivered when you need them. It's about hiring and paying professionals and suppliers. It's about working smarter to get more of what you want in your new home. So let's get started!

What Suppliers Do You Need?

Many folks will be helping you to build your home. Those who actually work on the house are contractors and laborers, discussed in Chapters 12 and 14, respectively. Those who furnish advice, other services, or materials are suppliers, and are covered in this chapter.

No matter what type of home you're building, you'll have lots of suppliers who don't actually build the house, but supply those who do. For example:

- ◆ Suppliers for conventional housing are those who help design and engineer the structure as well as those who furnish materials.
- ◆ Kit and log home suppliers include those who design, precut, package, and deliver the kit components.
- ◆ Manufactured home suppliers include the sales office and the suppliers of any materials used in the foundation and driveway, site-built garage, landscaping, and other add-ons.
- ◆ Modular house suppliers include the designer and the factory, as well as those who furnish advice or materials for erection.

In general, the most cost-effective way to find and hire good suppliers is to …

1. Find a large pool of potential suppliers.

2. Choose the most qualified from that pool.

3. Negotiate the best terms you can.

4. Hire the most qualified suppliers for the fairest terms.

However, finding a large pool of qualified suppliers and asking the most telling questions can be easier said than done. This chapter offers some specific ideas.

Drafting a Good Architect

Considering hiring an architect to help you design your own home? Scary, huh? Actually, no matter what home you build, a professional architect has probably already been involved in its design. Even a home kit was designed by someone, probably an architect. And you are hiring that architect as you buy that kit. It's the same for plan services and manufactured homes. Most homes built today don't directly hire the services of an architect. If your home is custom, you may want to hire an architect—at least to look over your designs and resolve any construction issues before they become problems.

On the Level

Professional architects are often members of the American Institute of Architects, or AIA. Check them out online at www.aiaonline.com for additional information and local referrals.

Architects offer a wide variety of services. They can do everything from reviewing and critiquing a plan service design to acting as the construction manager. Many specialize. Start looking for architects in the local telephone books. Also ask real estate agents, lenders, and contractors you've met. You don't want to find out that you hired the fourth-best architect!

How can you hire the best architect to design *your* home? Here are 20 questions you should ask any architectural firm you consider hiring. Of course, modify them when interviewing plan services, as well as kit and modular home manufacturers:

- Does your firm have a state architect's license?
- What national and local architectural groups are you affiliated with?
- How many people work at your firm, and how many of them are licensed architects?
- How long have you been in business?
- How much of your architectural experience involves owner-built or custom homes?
- What are some of the challenges you've overcome in planning custom homes?
- Can you furnish me with plan examples and references?
- How well have your construction estimates matched actual building costs?
- What services do you provide during the design, bidding, and construction phases?
- What are your fees for each of these services?
- Who in your firm actually does the work, and who oversees?
- Can you estimate construction costs?
- Can you recommend specific contractors, subcontractors, suppliers, and lenders?
- What additional services and fees will my project require?
- Can you meet the proposed design and construction schedule?
- How do you calculate your fees?
- How will design changes be made and charged?

- Will the plan require an engineer? If so, do you have recommendations and what are their fees?
- If a kit or manufactured house, how complete are the plans, and how specific are they to building codes and requirements in my area?
- Do you validate parking?

Calculating a Qualified Engineer

Does your house require an engineer? Local building code, the building site, or the house's design may require engineers. Most new homes don't. For example, local building code may require that an approved truss builder must engineer and manufacture the roof systems. Or your building site may have drainage or soil issues that you must resolve before construction. Or your design requires engineering for load.

Engineers who are hired for residential construction include the following:

- General engineering contractors (see Chapter 12)
- Structural engineers
- Electrical engineers
- Environmental engineers
- Consulting engineers
- Foundation engineers
- Geotechnical (soil, geologic) engineers
- Civil (roadways, bridges) engineers
- Sanitary engineers
- Water supply engineers
- Inspecting engineers

In most cases, you won't directly hire engineers, unless you're acting as your own contractor. Your architect, designer, contractor, subcontractor, or other professional will hire or at least recommend them to solve a specific problem.

Even so, you should be involved in the hiring process. It's *your* house and *your* money! At least know and meet any engineers hired by others. Let them know who they are really working for. Ask if you can call for progress reports or to get questions answered.

On the Level

Professional engineers are often members of the American Consulting Engineers Council (www.acec.org); American Society of Civil Engineers (www.asce.org); American Society of Heating, Refrigerating & Air Conditioning Engineers (www.ashrae.org); Associated Soil and Foundation Engineers (www.asfe.org); and similar organizations. Refer to Appendix B for more information.

Ka-ching!

You can often save money by directly hiring engineers, cutting out any referral fee that the engineers pay to whomever recommended and hired them. However, the engineer may charge you the same price either way. If you can't get a discount by hiring direct, let one of your other professionals do the hiring, and let them keep the fee.

Hiring a Materials Supplier

You'll need lots of materials to build your own home. The list will include concrete, lumber, sheathing, fasteners, doors and windows, fixtures, drywall, cabinets, roofing, and lots more. Where are you going to get all this stuff when you need it and at the lowest price?

First, review your Home Book for names of material suppliers that you've discovered or heard about. Maybe a contractor recommended them, or you've seen local ads that looked like their pricing is lower than others. Make sure they are first quality and not B or seconds. Select the three or four best suppliers for your primary materials (lumber, doors, drywall, roofing, and so on). Base your decision partially on whether the local professional builders use the supplier. Builders must get consistent quality at a low price to stay in business, so trust their judgment.

Then contact each supplier about bidding on your project. As with getting bids from contractors, suppliers will need specs and plans. Most important, make sure they get a comprehensive materials list, if available. Many architects and plan services will provide this list with plans. (But don't trust them to be complete.)

Most building material suppliers can develop a comprehensive materials list from your plans. They used to do it the old-fashioned way with paper and pencil. Today, a computer spits out pages of material requirements from a plan. However, remember the rule of computers: Garbage in, garbage out. If the plans aren't complete or the data isn't entered correctly, the materials list will be inaccurate.

You will probably have more than one materials supplier. The concrete, for example, may come from another supplier. The same rules of construction consumerism apply. Know what you're buying, and get fair prices from the best available suppliers.

Before selecting your primary materials supplier, interview one of the salespeople in charge of large accounts. You want to know about quality, pricing, discounts, delivery, terms, and advice. Ask the following questions:

- Are preferred brands (if any) readily available from the supplier? If not, can you easily order them?
- What are your prices on (list a few specific products such as a door model or roofing material)?
- What discounts and terms are available to licensed contractors?
- Do you offer the same discounts and terms to owner-builders?
- What discounts are available to owner-builders for cash-and-carry?
- How can I set up a home account (medium-size projects) or construction account (entire homes) with you, and what are the terms?
- When do you receive ordered materials from your primary warehouse?
- How soon can you typically deliver materials to our building site?
- How soon before the delivery truck leaves can I call in an additional order for delivery?
- Can you get me better pricing for concrete and other bulk materials than I can get myself?
- Will I have a specific salesperson assigned to my account (preferred)?
- Who would I talk to if I can't come to an agreement with a salesperson?

Ka-ching!

If possible, purchase lumber after the season's first rain. Wholesalers then consider the building season over and begin discounting prices to dealers and builders.

Keeping Your Suppliers Working Hard

Personally, you may or may not like a specific supplier. And it can be mutual. As long as you *trust* the supplier to do what is promised, that's really all you need. Your supplier says that 40 sheets of $4' \times 12' \times \frac{3}{8}"$ drywall are on the truck and will be delivered within the hour. It had better be there. No excuses. No sending 30 sheets. If he were sending 30 sheets, he should have called and told you before the truck was loaded. Why? Trust! It's an important part of all business relationships. In fact, it may be at the top of your requirements list as you look for contractors and suppliers. Do I trust this person to do what she or he says?

How can you keep your trustworthy suppliers working hard for you? Feed them! And what do suppliers eat? Money!

Construction may once have been their passion. But it eventually becomes a job. They do it because they know it—and it pays the bills. Most continue to enjoy their side of construction, but it is still how they feed their families and/or habits.

The point is—money, your money (technically, probably the lender's money)—is primarily what feeds their desire to help you. They work, you pay, they eat. It makes sense. So, remember the Golden Rule: Those with the gold, rule. Meaning you can use gold (money) to rule (manage) your suppliers. Here's how:

- Make sure all agreements keep work ahead of the money; that is, appropriate payment is made *after* work is done.

- Use incentives as needed to keep construction moving along the critical (most important) path.

- Make sure that delays cost contractors and suppliers more than they cost you.

- Spend a dime to save a dollar.

- Keep contractors and suppliers informed about draws and payments—especially if they're going to be late.

- Be willing to forego a discount to get speedy delivery on material needed in a hurry.

- Treat all suppliers as people who really want to do the best job they can.

How will you pay suppliers? That's where it's important that you and/or your contractor(s) keep good records. When were items ordered? When did they arrive? Were they what you ordered and the correct quantity? Who received and inspected them? Who needs them and when? Keep track of all the details.

When is the best time to pay a bill? If you earn a discount, pay it prior to the deadline. If there is no deadline, consider paying it immediately if your budget allows. You won't earn additional interest on the money, but you will earn the respect of your suppliers. You will be building trust. And trust is like money in the bank. You never know when you'll have to draw on it for an emergency.

Congratulations! You've hired a crew of contractors, suppliers, and advisors to participate in the building of your home. You're ready to start the actual construction!

> **On the Level**
>
> If you are your own construction manager, consider using a computer and spreadsheets (in a program such as Excel) to keep track of materials and orders. Yes, it's more work, but it pays off in the end because spreadsheets instantly calculate totals, update calendars, and keep your house on track.

The Least You Need to Know

- Find the best architect and/or engineer for your project by asking the right questions.

- The steady flow of materials to your job site is critical to its success.

- Motivate your suppliers to give you their best efforts.

- Hire only those people you trust with your money, and work toward building good relationships with them.

Part 4

Constructing Your Home

Someone's going to sweat. If you're building your own home, it's you. If you're hiring contractors and subs, it's them—while you perspire.

Actually, this final part will help keep you from sweating or perspiring. It's a comprehensive look at exactly what goes on at the building site, from marking the house corners to moving in. You'll know what your contractors are doing, and you'll know enough to participate in the process. You'll even pick up some handy tips passed along by experienced builders to make the job easier.

No sweat!

Chapter 16

What Goes Where?: Planning Construction

In This Chapter

◆ Why planning is vital to building your own home

◆ The importance of documenting construction

◆ How to prevent problems with contractors, subs, inspectors, suppliers, and lenders

◆ Working cooperatively with building inspectors

◆ Rules for making your building site safe

It's a big task, building your own home. Whether you're doing everything yourself or supervising construction, it will take hundreds of hours and thousands of dollars. It will take extensive planning to ensure that you get what you pay for—and what you wanted.

This chapter begins your path to a new home. It covers arguably the most important step: the planning. It guides you through working with contractors, inspectors, and lenders to get the job done right. Home Book in hand, let's start planning the construction of your own home.

Plan Hard, Work Easy

If you're like most owner-builders, you'll probably spend as much time planning your house as building it. Good for you! Time planning saves time building.

Planning what? As you learned in Chapter 9, 20 major processes go into building your own home, with each process including hundreds of steps. The more efficiently you can plan those steps, the easier, faster, and less costly construction will be. To review, these processes include the following:

◆ Preconstruction preparation

◆ Excavation

◆ Pest control

◆ Concrete

- Waterproofing
- Framing
- Roofing
- Plumbing
- Electrical
- HVAC
- Siding and/or masonry
- Barriers (doors and windows)
- Insulation
- Drywall
- Trim
- Painting
- Cabinetry
- Flooring and tile
- Gutters and downspouts
- Landscaping and driveways

Knowing what the processes and steps are will help you to manage the critical path. That is, you'll know when to schedule a contractor or order materials for the next process. Professional contractors agree that managing the critical path is one of the most important aspects of their job. And it's why owner-builders typically take twice as long to build a house as professional contractors.

The first half of this book covered most of the topics of preconstruction preparation and gave you an overview of the other 19 processes. This second half gets more specific so that you are comfortable supervising, managing, or actually building your own home. It won't make you a plumber; you'll need additional training and experience in residential plumbing for that. But it will show you what a plumber does and, using detailed plans, will guide you in the process.

So What Could Go Wrong?

It's an ugly sight, seeing a weathered, partially finished house. Someone started building his or her dream home and, for some reason, didn't finish it. Here are the primary reasons why people don't complete their homes. Learn from them!

On the Level

You're going to be dealing with lots of money: your money and your lender's money. Your lender can help you set up a construction banking account from which you will pay contractors and suppliers. Make sure this account is separate from your personal banking account so that you can keep clear track of costs and balances.

- Excessive design changes drive costs over budget.
- The contractor or construction manager lets building get out of control by not organizing well.
- Theft, vandalism, injuries, or natural catastrophes aren't covered by insurance.
- Contractors are paid before full inspection of their work.
- Sloppy banking pays some bills twice and others not at all.
- Owner optimism exceeds skills or resources.
- Conflict with spouse or partner stops funding or building.

Documenting Construction

Plan on having a still camera and, if possible, a video camera available during the construction of your home. The best choice is a small, easy-to-use digital camera. You can purchase them for under $100. One owner-builder carried a small automatic camera in his tool belt for daily use, and brought a video camera to the site each weekend to chronicle construction.

You don't need a computer to take advantage of digital photography. You can take the storage disk from the camera to photo service stores and booths to have selected shots turned into prints.

I can give you many good reasons why you should chronicle the construction of your new home. First, photos will make a great coffee-table book, showing the lot and house as it takes form. Second, photos will document for lenders, building departments, and others the progress made in construction. Third, photos or a videotape can document problems met along the way, including shoddy work by a contractor, incorrect building material delivered, or other problems.

Of course, continue using your Home Book to keep daily notes, document problems and concerns, and as a reference book that includes the names and telephone numbers of inspectors, contractors, and suppliers.

Ka-ching!

For less than $20, you can get a few hundred business cards printed up with your name, the building site location, your contractor's or advisor's name and phone numbers (if appropriate), your lender's name and phone numbers, and any other information you should share with others. Include wired phones, cell phones, mobile phones, faxes, and any other phone numbers. Letting people know how to contact you and the other decision-makers in your construction will make things go more smoothly.

Contact!

Managing a construction site means staying in contact with everyone: contractors, subs, tradespeople, inspectors, lenders, family, and others. Thank goodness for cell phones! Get one. Use it. Share your number.

If you are at work during the day and your boss allows it, include your work telephone number. And if you have an e-mail address, include that, too. If that's too much info for a 2" × 3½" card, go to 4" × 5" index cards or whatever it takes.

Murphy, the Contractor

Murphy's Law states: Whatever can go wrong, will, at the worst possible time. Murphy has obviously built his own home. Needed subs will not show up. Suppliers will send the wrong materials. Vital equipment will break. An inspector will have a fight with his spouse and will take it out on you.

So how do smart general contractors keep Murphy's Law off the construction site? Here are some tips:

♦ Each evening, plan for the next day's work.

♦ Communicate. Let contractors, suppliers, laborers, inspectors, lenders, and other partners know what is expected of them and confirm when they will comply.

♦ List who is expected at the building site (contractor, inspector, lender) and be ready to help them.

♦ Make sure the needed materials will arrive prior to the day they are needed. If the contractor picks up the material, call to remind or confirm.

♦ Do not allow alcohol or other mind-altering substances at your construction site.

- Don't work or allow others to work at your building site when they are sick, excessively tired, or when there is inadequate lighting.

- Keep the work area clean of debris, especially lumber-trim pieces.

- Make sure that power, water, and sanitation (outhouse) are available at the job site.

- If you don't have a pickup truck for hauling things around, buy or rent a utility trailer. An open one is cheaper, but a closed trailer can also serve as your office or tool storage.

Code Red _____

Beware of buyer's remorse! It's a common malady that strikes folks who spend more money than they have in their pocket. The symptoms are getting nervous about previously made clear decisions. "What the heck am I doing, building my own home? What was I thinking?" The solution is to know that it will strike you at least once during the building process. Remind yourself why you are building and how you thoughtfully and carefully came to the decision. Review your Home Book and talk with others who have supported your decision. You'll get over it.

Need some more real-world laws from the construction site? Try these:

- It is impossible to make anything totally foolproof because fools are ingenious.

- Everything takes longer than you think it will.

- Multiply all delivery estimates by two.

- Whatever you buy this week will be on sale next week.

Changes happen. During construction, you or your general contractor may determine that a design change is needed. How can you manage changes?

Chapter 11 covers change orders and how to keep them from stopping the construction of your home. In essence, make sure any changes are documented and that all folks who need to know (contractors, subs, lenders) know about it.

Murphy, the Sub

As you've learned, subcontractors are specialists rather than generalists. They typically specialize in one or two of the numerous construction trades. You will rarely meet one who is trained for both plumbing and electrical work, for example.

So subs look at your house a little differently. They see a component in detail rather than the whole picture. And they are primarily interested in getting in and getting out as quickly as possible so they can move on to the next job.

Whether you are your own subcontractor, the construction manager, or just a smart owner, building can be a challenge of communication. That's where clear building plans are vital. Make sure your architect or plan service provides clear, detailed drawings that tradespeople can read and use. Also, consider setting up an initial meeting with each sub or tradesperson to review the plans together and answer any questions. Make sure that any agreed-upon changes are documented in writing.

Thankfully, you've done your homework in selecting and hiring subs, so you know whether your partner is a good communicator or not. Experienced tradespeople will supply their own tools, including specialized trade tools. If they don't, you didn't hire the right people. Even so, some contractors may need to rent or buy a special tool they would only use for your job. Hopefully, their bid includes getting the tool. Depending on the tool and cost, you may want to rent or buy it. If it saves more money than it costs, get it.

The biggest challenge in working with subs and tradespeople, say experienced contractors, is getting them to show up when they say they will. Sometimes it's due to overbooking, but often it's not their fault. A job at another house takes longer than scheduled. Equipment breaks and they must first repair it. Inclement weather slows down prior projects. Illness strikes the crew. Murphy rules!

I've known subcontractors who will "start" a job on time by placing some of their materials or tools on the job site, but not show up to *really* start the job for another week or two!

Some subs are notoriously behind schedule. It's typically those whom you need right now to keep things going. Hopefully, you've identified which subs those are in your area and either have not hired them or have added some extra time to your schedule for them.

On the Level

If you rent equipment or tools, make sure they are covered by your builder's risk-insurance policy. If not, purchase additional coverage from the rental service to cover loss, damage, and theft. You want to rent the equipment, not buy it.

Ka-ching!

Call your subs the evening before they are scheduled to be on your site to confirm arrival time. And get their cell phone number in case they don't show up. Make sure they have your cell number, too.

Murphy, the Inspector

Building inspectors are powerful. They can stop construction at your building site, putting everything on hold. Fortunately, that rarely happens. Most building inspectors are hard-working folks who have a job to do.

What's their job? Building inspectors inspect the structural quality and general safety of buildings. In larger municipalities, they specialize. Some handle residential construction while others inspect commercial buildings or highways and bridges. Within residential inspectors, some check electrical while others inspect plumbing or HVAC systems. So you may work with more than one inspector during the construction of your house.

Before any inspections occur, you or your contractor must file for a building permit (see Chapter 11). Part of that process requires a plan examiner to review your building plans, making sure it is designed to meet local building codes. Once approved, the inspector's job is to verify that the structure is built according to the approved plans.

If you've purchased a manufactured home, then it was constructed using HUD or state building codes, so local building inspectors will only check how it is installed and hooked up. Typically, if any structural changes are made to the manufactured home on site (such as modifying a roof attaching a garage), you must involve a state inspector.

Once you or the contractor is ready for a required inspection, call the building department to schedule it. It's best if the construction manager (you or someone else) is on site as the inspection is done. Otherwise, questions or problems may delay the construction process.

Building inspections are typically done on the foundation, underflooring, rough-in, drywall, and final stages. A building inspector will fill out an *Official Notice of Inspection* form including the date, inspector's name, permit number, owner's or contractor's name, construction site address, and a list of corrections to be made.

For example, a final inspection may require the following:

- Strap weatherhead mast to wall
- Label breakers in service panel
- Wrap gas pipe within 6" of grade
- Paint all exposed exterior gas pipe
- Provide anti-siphon valves on all exterior hose bibs
- Provide stairs to all exits
- Slope grade away from house for 5 feet

Finally, the inspection notice will include a deadline such as this: "A reinspection shall be required to correct items within 180 days." What can go wrong with inspections? Noncompliance. That means someone didn't correct previously listed problems. So, your job, or the contractor's job, is to make sure that everything on the list gets done. That's why it's often most efficient to meet the inspector at the building site and listen. Don't argue! Building inspectors are like baseball umpires. Ask how you can make sure that you or your contractors comply with the notice.

Also remember to ask for the inspector's business card and ask if contractors or tradespeople can call with additional questions. Many inspectors attempt to be helpful. It's not a power struggle. Their job is to make sure that the structure complies with the approved plans. Help them do their job, and they will typically help you do yours.

Murphy, the Supplier

Smart construction managers know that not having needed materials on site can be costly. Expensive people don't like to wait. So, hiring dependable materials suppliers is crucial to building your home on time and on budget. Saving $100 on concrete, for example, is false savings if a crew must stand around for half a day waiting for delivery.

Ka-ching!

It's a good idea to have a backup supplier. If your job can't go ahead without a specific material, it may be less costly to order it from another supplier at a higher cost than to have people standing around while you pay them.

The solution? Communication! First, make sure that you know the delivery reputation of suppliers before you contract with them. They don't have to be faster than fast. They just need to be dependable. For example, a supplier who needs a week for delivery simply needs a week's notice. That's the construction manager's job: to place the order on time to get delivery on time.

Also, make sure that everyone is clear on who orders what materials. If your foundation contractor shows up assuming that you've ordered the concrete truck, and you assume that he has, well ....

Fortunately, technology offers voice phones, answering systems, cell phones, faxes, e-mail, and other cool tools. However, it still takes people to communicate!

Murphy, the Lender

What can go wrong with your lender? You name it! Draw inspections can fail. Paperwork can get lost. Messages are mislaid. You lose your job. Lots of stuff. However, most of these potential problems can be minimized with good planning. Knowing what *can* happen makes you plan smarter.

The most common problem that can occur with lenders is that a draw payment gets delayed. That means a specific inspection didn't pass the lender's requirements. So, the obvious solution is to know what those requirements are and to make sure that you meet them. Keep your lender informed about progress. Offer photos if possible. Find out who actually makes the lender's inspections and keep in touch.

Job Site Rules

Depending on how you participate in the construction process, the job site will be your constant source of worry, or your home away from home—or both. You can make the job site easier to manage by establishing a few basic rules:

- Designate your construction office. It may be a small shed, a sawhorse bench in the garage, or an on-site travel trailer. It's where you store permits, plans, inspections, and other paperwork. It may also secure expensive tools and materials.

- Make sure everyone who works on the site knows where the first-aid kit is located, and never keep it locked up while the crew is on site.

- Establish temporary utilities. Depending on the site, make sure you have a temporary power pole with weatherhead, meter, service panel, and receptacles. Also, get water to the site, if needed, for concrete or fire protection.

- Make sure the site is as safe as possible. Quickly remove building debris, or make sure contractors do so before it becomes a hazard.

- Protect building materials. Make sure that all materials stored at the site are safe from the weather and secure from vandalism. Losing materials can cost both money and time.

- Provide security. Depending on the risks where you are building, hire a security service or even a neighbor to watch the site when no one is working there.

- Make sure everyone works safely. Because it is your building site and you are ultimately responsible, you have the right to require safe working habits and equipment.

- Provide sanitation facilities for workers.

- Post No Trespassing signs at the property edge and include a telephone number for reports.

- Fill in or cover holes and ditches as quickly as possible.

- Make sure you have builder's risk insurance (see Chapter 13).

- Verify that all contractors, tradespeople, and laborers working at your site are covered by worker's compensation insurance (also covered in Chapter 13).

- Check all scaffolding or ladders used at the building site for safety and sturdiness.

- Make sure that there is sufficient clearance from all power lines.

- Set a good example of the behavior you want on your building site.

> **CAUTION**
>
> **Code Red**
>
> Construction sites are kid magnets. They draw neighborhood children while you're working as well as when you're not. If found at the site, show them where they can safely stand (off the property) to watch, and ask them to tell you if they see anyone on the site while the crew isn't there. Reward good behavior with a supervised tour.

We'll cover more specific job-site safety rules in the next chapter.

The Least You Need to Know

◆ Smart planning is the key to successful building.

◆ Early on, set up a construction bank account and begin documenting what occurs at your building site.

◆ The more time you spend on planning, the more time you'll save on building.

◆ Guard your building site against injury and financial loss by establishing some basic safety rules.

Chapter 17

On Your Mark: Site Preparation

In This Chapter

- ◆ Working safely at your building site
- ◆ Where to store building materials
- ◆ Surveying and clearing the building lot
- ◆ Locating the structure and planning utilities
- ◆ Finding valuable resources

This chapter starts with a class in construction-site safety. Once you pass the test, you can begin surveying and clearing the building site, and then marking the location of structures and utilities.

Whether you are your own construction manager, the entire labor force, or a sidewalk supervisor (once the sidewalk is in!), this chapter offers clear directions for getting your building site ready for construction. Grab your hard hat and let's get to work.

Safety

The construction site is a dangerous place to be, especially if this is the first one you've been on. There are hundreds of perils that can remove fingers, break bones, leak blood, and other ghastly things that require a visit to the ER. You certainly don't want to be injured, nor do you want anyone who works at your construction site to be. So listen up as we cover the basic safety rules of construction:

- ◆ Relax; getting frustrated or angry can quickly lead to distractions and injury.
- ◆ No alcohol or drugs are allowed on the building site. Period.
- ◆ Watch out for sunstroke, heat exhaustion, and sunburn.
- ◆ Always use protective goggles when using power tools.
- ◆ Never use electric power tools while standing in water.
- ◆ Unplug power tools before changing a blade or drill.
- ◆ Be careful of saw blades that get stuck in wood and reverse direction or kick back.
- ◆ Make sure you know the safety rules for each power tool you use; check the owner's manual if you're unsure.

- Don't try to cut off more than your power tool can chew.
- Wear heavy-soled work boots, preferably with steel toes.
- Use rubber-soled work boots for traction when working on the roof.
- Wear gloves when handling lumber, metal, or other materials that can injure your hands.
- Wear a hard hat when working under or near overhead workers.
- Wear a protective breathing mask when handling or installing fiberglass insulation and when sanding.
- Cover or tie back long hair.
- Remove protruding nails from lumber *before* they impale your foot.
- Place all scrap lumber in a large container for reuse.
- Place all nonrecyclable construction debris in a separate container for the dump.
- Select the appropriate tool for the job; don't try to use a tool for something it wasn't designed to do.
- Make sure the cutting edge of the tool is pointed away from you.
- Don't leave tools on ladders, scaffolding, or other high areas.
- Don't carry sharp tools in your pockets.
- Don't distract someone who is using a power tool.
- When lifting, stand close to the load, bend and lift with your knees rather than your back.
- Don't twist or turn while lifting or carrying a heavy load.
- If you are higher than five feet up a ladder, make sure that someone or something is stabilizing it.
- Make sure you know where the first-aid kit is and that it is fully stocked. Bandages are very expensive at the ER.
- Keep your eyes open for workers who aren't as safety conscious as you are.
- Don't work when you're exhausted, especially around power tools or on high places.
- Stay alert and keep your mind on your work!

Okay, here's the safety test. If, once you've completed building your house, you have the same number of members in the same condition as when you began, then you pass. Otherwise, you fail.

Storing Materials

Over the course of constructing your house, you'll need to store some materials for days or even weeks. Where? Well, once your garage is built and enclosed, you can store materials and tools inside. But that's later. For now, here are some material storage options:

- Coordinate deliveries so that needed materials aren't on site very long before they are used.
- Rent a lockable storage container or trailer that is parked on the site during construction.
- Place plywood and sheathing horizontally on pallets or on skids of dimensional lumber to keep it from contacting soil.
- Stack lumber flat and keep it bundled until needed.
- Store bricks and blocks on pallets placed near where they will be used.
- Keep shingles bundled or wrapped until installed.
- Shingles can be delivered to and stored on the upper half of the roof (installation begins on the bottom edge).

Ka-ching!

If you are building in an area where theft or vandalism is likely, hire a security service to watch the site during times of high materials inventory.

◆ Once the house is enclosed and barriers are in place, you can store materials inside.

◆ Use your vehicle or materials trailer to store smaller tools and materials.

Remember, at any given time you may have many thousands of dollars in materials on your job site. Theft or damage can cost not only replacement, but also lost construction time.

Surveying

I can see that you're getting anxious to start working on your building site. So here's your first job: surveying. You can do it yourself or hire someone to do it. Most folks hire a surveyor for a few hundred dollars to make sure the property line is accurately identified. This is especially important on smaller rural lots. New suburban lots have already been surveyed, so it's just a matter of finding and clearly marking the corners.

A plot plan is a scaled plan showing property boundaries and the building-department-approved location of the house, garage, driveway, and setbacks (the distances that structures are set back from boundaries). In Chapter 11, you learned about building permits, and you filed a plot plan for your house. Before any construction begins, make sure that you identify and mark all corners of the building lot.

A survey of your property measures the width, depth, and height (elevation) of the parcel. The builder's level is a surveying tool, much like a telescope with a leveling vial or other component to make sure the telescope is level. The builder's level has a circle with the degrees marked on it. It's all mounted on a tripod. There's also a target pole, a tall pole with measurement numbers written on one side.

Code Red _____

If the building department determines that you have built your house within the designated setback, you may be required to move the structure. Ouch! Probably, though, they will require that you file for a variance, a time-consuming and sometimes expensive hassle.

Typical plot plan for a residence, garage, and deck.

The object of surveying is to look through the level telescope at the specified distance and angle, and read numbers on the target rod. Here's how lines are surveyed:

◆ The builder's level is placed directly over a known marker, such as a corner stake.

◆ The telescope is leveled and pointed in the required direction, measured in degrees (360 degrees in a circle), minutes (60 minutes in a degree), and seconds (yep, 60 seconds in a minute).

◆ A measuring tape is run from the known marker for the specified distance in the direction that the telescope is looking.

◆ The rod is then held vertically at the specified distance so it can be seen through the crosshairs of the telescope. The numbers on the target rod indicate the elevation of the new location.

◆ The new site, at the designated distance and angle from the known marker, is marked and the process begins again to find the next marker.

The builder's level is aimed to view the target or ground rod.

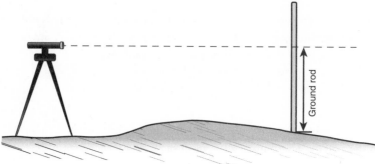

Surveying is done to mark property boundaries, find the elevations of a parcel, and locate the exact building site for the primary structure and any roadways. If you'll be including a septic system and drain field, you will need to mark these as well.

It's especially important that you survey and mark the corners of the proposed house location survey. It shows you, any contractors, the building department, and the lender where the house will be. It's a critical survey for smaller lots in which moving a house two feet in either direction for a better view will encroach on setbacks.

Besides the builder's level, other tools are available for surveying building sites. Another is called the level-transit. In addition to measuring horizontal angles, a level-transit can measure vertical angles up or down 45°. And these tools have high-tech variations, such as laser level transits.

Ka-ching!

Save yourself some time and money by making sure the corners of your building lot or parcel are clearly marked and flagged by the prior owner before buying it.

You can rent a builder's level or a level-transit at larger rental yards by the hour or day. Or you can purchase a basic level package including tripod and rod for under $300 through larger building materials suppliers. If available, buy a video on basic surveying at the same time. Or you can hire a surveyor or engineer to do the work for you. Depending on the size of your lot and the angles to your home, expect to pay from $200 to $1,000.

Your building site may require a topographical survey as well. If the lot isn't relatively level, you will need to move dirt for the foundation to level out the yard. Before construction, you'll probably need one of these surveys. Again, you can do it yourself or hire someone to do it. Depending on how many levels your building site has, it can get very expensive.

If you hire a surveyor, he or she can also calculate "cut and fill" for your site. It will show the excavators where to remove (cut) soil and where to add (fill) it. Hopefully they will balance, and you won't need to bring in or take out soil from the site.

Clearing

Most building sites are relatively clear, especially those in subdivisions. The subdivider has probably already removed anything that looks like it might hamper construction.

The next task in preparing your lot for construction is probably removing any existing structures, trees, and stumps. That typically requires some heavy equipment, or at least lots of perspiration. (If it looks like the developer scalped trees from your building lot, there may be a good reason. Heavy equipment attempting to level lots can damage the root systems of trees on the lot. The first bad storm may bring down the tree across your living room, so many builders err on the side of removing most or all trees from developments.)

Chances are that you'll hire someone to do this work. Smart move. Heavy equipment is expensive and you need training to use it. However, you can save some money by making sure you have a clear site map that indicates what needs to be removed and what stays. In fact, it's a good idea to clearly mark anything you want to stay, such as specific trees. Tie a yellow ribbon 'round that old oak tree! Or "tag" it with spray paint.

Also, go over the site map with the excavation contractor so that everyone is clear on what needs to happen. Also, if your contractor is not the equipment operator, ask to have the operator sit in on your discussions. Remember: Communication is the key to successful building.

So what does an excavation contractor typically do?

♦ Removes existing structures

♦ Removes tree stumps (you may need a tree service to actually remove any trees)

♦ Removes large rocks or boulders from the site

♦ Levels the building site, adding or removing soil as needed

♦ Removes soil as needed for the basement, crawl space, slab, and/or foundation

♦ Grades the building site for proper drainage

♦ Once the foundation is in and treated, backfills or replaces dirt against the foundation

Treating the foundation will come later. It typically means applying a sealant that keeps water from being absorbed and passed by the foundation. I'll cover that in the next chapter, along with excavation.

 Code Red

If possible, don't let the excavator bury tree stumps. They can rot below ground and become a home to termites, yellow jackets, and other pests. Have the stumps hauled away or, if allowed, burned on site.

Locating Structures

How can you identify the location of your structures? As mentioned earlier in this chapter, you'll be surveying and marking the perimeter of the structure with stakes and flags. Let's go a little further with that topic.

Depending on the size of the building lot and the detail on your preliminary site plan, you may be able to measure from the lines to mark the exact building site. For example, if the south edge of the structure begins 20 feet north of the south property line, it's an easy task to measure and mark the south edge. Then, if the sides are 10 feet in from each edge, you have two more perimeter lines. Finally, you refer to the site plan and measure in as appropriate for the structure's north edge.

Many sites, however, won't be that easy. They aren't rectangular or level or are too large to depend on measurements from lot lines, such as on acreage. Of course, the larger the site, the less exact it needs to be. The issue is to make sure you're not building where you aren't supposed to.

Code Red _____

Vandals and children may think it's fun to move marking stakes on a building site. Even a couple of feet can make an expensive difference. Always take photos of your staking and make a confirming measurement before committing to final construction.

How do you mark corners? Typically with surveyor's or gardener's stakes stuck in the ground at the appropriate point. To make sure they are marked well, you can use spray paint to mark the stake and surrounding ground. You can also tie colored ribbons on stakes to easily identify them.

If you need to verify a right angle, measure 3 feet from the corner along one edge and mark it. Then, from the same corner, measure 4 feet along the second edge and mark it. Finally, measure the distance between the two marks. It should be 5 feet. If not, adjust the angle of one of the two edges until the distance is 5 feet. Builders call this the _3-4-5 formula_.

Planning for Utilities

It's also important to identify and mark utilities as they enter the building site. That is, place a stake where the sewer line, water line, and underground power line are "stubbed" or enter the property.

How can you tell where they come in? The lot developer, utilities department, or building department can help you identify them. You can often verify them by using a metal detector. Paint the stake or the ribbon to indicate the type of service: W for water, E for electricity, G for gas, C for cable, and so on.

If your electricity and cable come in above ground, put a stake where they will enter the building site. The utility company can advise you. You will eventually install an electrical meter and box, and you'll want to know which corner of the house to install them on later.

What if you're not getting all your services from a utility service? What if you're digging your own water well or accessing a spring? What if you need to install a septic system? Fortunately, your plot plan will show the location of the well (because you'd never buy a lot without water available) and/or the septic system. The map will show the location of the septic tank and drain field, with measurements. Your excavation contractor will use the plans to prepare the site for the tank and field. But don't get in a hurry to install your septic system and drain field yet. Make sure that heavy equipment at your site won't drive over the field, destroying it.

On the Level _____

If your building site requires a well, first talk with a well drilling service or consultant to determine the best location, probable depth, and estimated cost. A 2,000 sq. ft. home typically requires about 10 gallons per minute (gpm) of water. If your well cannot produce at least 5 gpm, some local water departments will require that you install an auxiliary storage tank. Normal pressure should be 20 to 40 pounds per square inch (psi). Lower pressure may not be sufficient and higher pressure may require a pressure limiter to keep from damaging pipes and fixtures. Talk with an expert. Most folks install the well prior to installing the foundation, just to make sure the two are relatively close to each other.

To get electricity on the site for construction, you'll need to install a temporary power pole. It's typically a 4" × 4" × 12' post with a weatherhead (electrical line entry point), conduit, meter, service panel, electrical receptacles, and a metal grounding stake.

Most busy contractors have a few temp poles made up that they move to new jobs. The pole is placed at least 3 feet and the grounding stake a minimum of 2 feet into the ground. The incoming lines have drop loops that offer slack in the line, in case something moves. The covered service panel will include a main shutoff for power. The bare ground wire is attached to the ground stake with a bonding clamp.

Make sure you rent a portable toilet and have it placed on your building lot before construction begins (and before nature calls). You'll be glad you did.

Can't get the power company to drop a line to your building site soon enough? Consider paying a neighbor's monthly electrical bill in exchange for letting you run a heavy extension cord from their lot to yours. You can do the same deal if you need water—except use a hose.

If you are considering alternative power sources, check out my book *The Complete Idiot's Guide to Solar Power for Your Home* (see Appendix B), available through bookstores or at www.MulliganBooks.com.

Site Prep Resources

Rental yards are useful resources for constructing your home. It's often less expensive to rent equipment than to buy it for a single house.

What can you rent?

- Hand tools
- Power tools (electric and pneumatic)
- Heavy equipment
- Trailers
- Air compressors
- Backhoes
- Post hole diggers
- Excavators (full-sized and mini)
- Paving equipment
- Painting equipment and ladders
- Portable toilets
- Temporary power poles
- Pickup trucks
- Flatbed trucks
- Storage units
- Construction site lighting units
- Lifts and hoists
- Demolition tools
- Soil tampers and compactors
- Ladders and scaffolding
- Pumps
- Stuff you've never heard of and probably don't need

Check the local telephone book's business section under Rental Service Stores & Yards.

Another useful resource for building your own home is construction consultants. The telephone book files them under—you guessed it—Construction Consultants. Some specialize in advising owner-builders, so shop around. They typically charge by the hour or a percentage of the job's value. If you run into a problem you can't handle, calling up a construction consultant may save you thousands of hard-earned dollars.

Next, it's time to start actually excavating the site and installing the foundation. That's coming up in the following chapter. Dig you later!

The Least You Need to Know

◆ Working safely takes just a little more time and saves you lost time.

◆ Make sure you have a safe and secure place to store expensive building materials and tools.

◆ You can survey your own building lot or hire someone else to do it.

◆ Once the lot is clear, stake out the exact location of all structures, driveways, and utilities.

◆ You can rent through rental yards many of the tools that are too expensive to buy for one house.

Excavation and Foundation

In This Chapter

- ◆ Excavating your building site
- ◆ Framing the foundation
- ◆ Pouring concrete and building foundation walls
- ◆ Installing a septic system, if needed
- ◆ Finding resources for excavation and foundation

It doesn't look like a house yet, does it? The building site has been graded and a few flagged stakes are sticking out of the ground to mark the boundary and where the house will be someday. There's a temporary utility pole for electricity and the outhouse for workers. That's about it.

Take heart! You'll begin seeing more progress as the foundation is excavated or cut into the soil and then built. This chapter covers excavating the site for the foundation, working with concrete and masonry, building foundation walls and piers, pouring slabs, and other tasks.

Excavation and concrete work are some of the most labor-intensive tasks on your job site, so make sure that you work safely (following the rules outlined in Chapter 16). Grab a shovel and let's get to work.

Excavating for the Foundation

Houses require foundations. Why? Foundations support and stabilize the walls and roof. That's a pretty important job! Excavation is the removal of soil to make room for the building's foundation. So before excavating, let's learn more about foundations.

The foundation itself needs a foundation. Most houses are so heavy that the foundation must be widened at the base to keep it from sinking into the ground. The wide part of a foundation is called the footing. Its size depends on the kind of soil under it. Most footings are designed to carry 1,000 pounds per square foot. A two-story house will have a wider foundation footing than a single-story house. Typically, the footing is twice as wide as the wall. That is, the footing for an 8-inch foundation wall is usually 16 inches wide. However, don't go by this; go by your house's foundation plan.

The foundation wall can be just 1 foot tall or it can be, with support, 10 feet tall or more. Or it can be shorter on one side and taller on another. It just needs to be level on top where the floor will be installed, which I'll tell you about in the next chapter.

How tall and wide is the foundation for your home? Refer to the foundation plans approved by your building department. It will include the foundation's layout, dimensions, size, elevation, and height.

The most common foundation is the continuous wall. It can be built of stone, clay tile, block, brick, concrete, treated wood, metal, or other material. Reinforced concrete is the most popular. Continuous walls are used to support heavy loads or to ensure a crawl space or basement.

A slab foundation is a solid floor set directly on the soil. Most modern garage floors are concrete slabs with a continuous wall foundation around the perimeter.

Building Your Vocab

Knee or **pony walls** are wooden frames that vertically extend the foundation wall to create a level building surface on a sloping site. They are also known in the construction trade as under-floor buildup.

A step foundation is a continuous wall of variable height. A short wooden wall, called a *knee* or *pony wall*, is built on top of the step foundation to bring it up to a single level.

A pier foundation is a series of concrete piers and footings that support the structure. If made of pressure-treated wood, it's called a pole or post foundation.

It's important to know the type and size of the foundation now because it tells you what soil needs to be removed or excavated for the foundation. It's best not to remove any more compacted soil than necessary when excavating for a foundation. The more the earth has been disturbed, the more difficult it is to ensure that the foundation won't move in the future.

Still with me? Good! Let's take a look at the steps needed to excavate and build your home's foundation:

1. Lay out the shape of the foundation and footing.
2. Excavate soil to the proper depth.
3. Build the footing forms.
4. Add reinforcement if needed.
5. Pour the concrete footing into the forms.
6. Build the foundation forms.
7. Add reinforcement if needed.
8. Pour the concrete foundation into the forms.
9. Finish the concrete and install anchors.
10. Remove the forms.
11. Waterproof the foundation and add drains.
12. Backfill or replace soil against the foundation to the grade.

Laying Out the Foundation

Before beginning excavation of soil, you need to know exactly where the foundation will be. Chapter 17 introduced marking the perimeter of the foundation with stakes. Now let's begin laying out the foundation.

The proven way of marking a foundation location is to build reference frames, called batter boards, a few feet outside of the foundation outline. These become reference points for excavation. Strings stretched between the batter boards actually outline the foundation. In addition, the batter boards mark the height or elevation of the foundation top. Pretty handy!

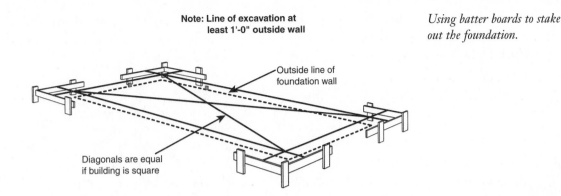

Note: Line of excavation at
least 1'-0" outside wall

Outside line of
foundation wall

Diagonals are equal
if building is square

*Using batter boards to stake
out the foundation.*

Once you outline the foundation location, you can begin excavation. There are many ways to remove soil; your choice depends on what type and size of foundation you're installing, at what depth and elevation you're installing it, and on the soil itself. Short foundations on level lots may simply have a front-loader scrape away soil to a specific level and build forms on top. Deeper foundations require trenching and more work with the soil.

You can see what tools you'll need for excavating and building the foundation:

- ◆ Shovels
- ◆ Backhoe or trencher
- ◆ Concrete and mortar mixers (or delivered concrete)

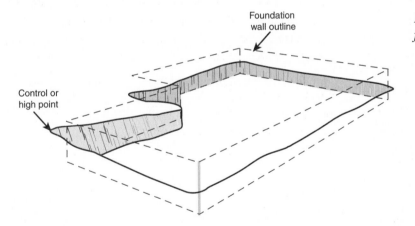

Foundation
wall outline

Control or
high point

*Excavating for the
foundation.*

Framing the Foundation

Forms are typically built of dimensional lumber (2" × 6", and so on) and plywood panels, depending on the size of the form. Most foundation contractors have pre-built forms they reuse.

Special forms are available for making piers or round support columns. Some are made of metal while others are simply cardboard tubes that can hold cement until cured, and then removed.

Foundation forms sometimes need to have openings in them for basement windows, pipes, utilities, or other uses. These spaces are called block-outs or bucks and must be installed with the foundation framing. For example, a foundation that requires a notch for holding a wooden beam will need a buck installed on the inside of the framing at the location specified in the foundation plan. Don't forget!

Foundation forms are typically held in place in two ways. The outside is supported by a brace and stake to keep the form from moving. The inside spacing is retained with a wooden spreader, tie rod, or metal strap—anything to maintain the correct spacing while not getting in the way of the concrete.

Framing for a poured concrete continuous wall foundation.

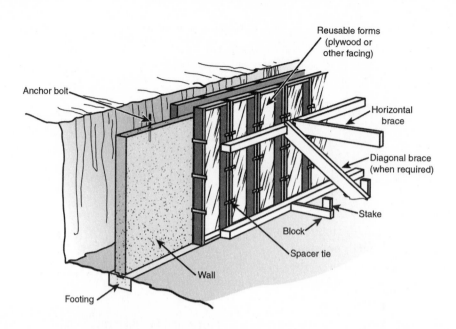

Reusable forms (plywood or other facing)

Anchor bolt

Horizontal brace

Diagonal brace (when required)

Stake

Block

Spacer tie

Wall

Footing

Ka-ching!

You can also rent one-piece forms that are tapered so that you can pour the footing and wall in a single step. They are built in sections and use special fasteners to hold the walls in place while concrete is poured and dries or cures.

Depending on the load and width of the foundation walls, concrete usually requires reinforcement to keep it from cracking and weakening. Check your approved foundation plans. Reinforcement is typically in the form of reinforcement bar or rebar—lengths of round steel bars that are installed in the form where they will be surrounded by wet concrete to add stability when the concrete dries. For some smaller installations you can use heavy wire mesh. Rebar is spliced into longer pieces by overlapping two rebars by 8 to 30 inches (based on the rebar diameter) and wrapping it with wire.

Rebar is identified by a number: #2 is ¼ inch in diameter and #8 is 1-inch rebar. Special supports are available that hold the rebar in position. They are called *bolsters* and *high chairs*.

Mixing Concrete

Once the forms and any needed rebar are in place, it's time to pour concrete. Plans may call for concrete footing or piers and a concrete, block, or wooden wall.

You can buy ready-mix concrete by the bag and rent a portable concrete mixer to do the job. However, most contractors and many owner-builders order delivered job-mix concrete. What you choose depends on your budget, your physical endurance, and your building site. It's hard work.

Concrete is a mixture of cement, sand, and rock mixed with water. Numerous types of Portland cement are used in concrete depending on whether strength, fast curing, or freeze-resistance is needed. Sand is also called fine aggregate, and gravel or crushed stone is called coarse aggregate. A common ratio is 1:2:3, meaning one part cement, two parts sand, and three parts rock.

Add-mixtures are also available to improve workability, remove air bubbles, or accelerate setting and hardening. Make sure you know what you're doing before using add-mixtures, or you could wind up with an unusable foundation.

When mixing concrete, how much *clean* water should you use? Enough! That typically means about four to six gallons, but can be more or less depending on the intended thickness and the dryness of the sand. Of course, too much water weakens strength.

It's best to leave it to the foundation contractor and/or your concrete contractor to select the appropriate concrete mix for your job. If you're a real do-it-yourselfer, you can do the research and the work yourself, but the foundation is one place you don't want errors, so consider getting expert advice.

Also, if you're mixing your own concrete, remember to test it before pouring an entire foundation. Mix a small batch or two and see how well it mixes and how long it takes to dry.

On the Level

A typical bag of ready-mix concrete includes the cement, sand, and rock needed for 1 cubic foot (c.f.) of cement (1' × 1' × 1'). (Smaller bags are available.) Just add water. The bag's contents weigh about 93 pounds!

Pouring Concrete

Pouring or placing concrete should start within about 20 minutes of mixing it. Why? Because, in warm weather, initial setting of the concrete starts by then. Of course, that time will vary depending on the outdoor temperature. Hotter means it will dry faster. Transit-mixed concrete suppliers can mix in additives to speed up or slow down the set.

What should you do with concrete that begins setting up too quickly? Discard it!

When pouring concrete into forms, use a concrete spade or vibrator to make sure the concrete doesn't retain air pockets. You can fill small pockets on the outside edges of walls with mortar or neat (without sand) cement, but internal ones weaken the foundation. Don't stir up the concrete mush very much, as finer materials will come to the top.

If you're pouring a concrete slab for a garage or walkway, you'll need to smooth the top side of the concrete, called finishing. This is done with special finishing tools such as a straightedge, edgers, trowels, floats, and a finishing broom. Finishing concrete is a craft. If you don't have experience with it, invest in some experienced help.

Concrete must dry, or cure, before it is useable. Because flat surfaces will dry much faster than the interior, keep them wet for three to five days. You can also add special additives to concrete that help it cure consistently. Ask for professional advice from your concrete supplier or contractor.

Code Red

Never pour concrete on top of the vapor barrier without a minimum of 2 inches of sand below the concrete. Why? Because concrete sets best when it dries equally from above and below. Uneven drying develops small cracks, called spider cracks, in concrete. Water can later get into these cracks and cause problems.

Building Block Walls

Instead of poured concrete foundation walls, you can opt for concrete blocks. They are popular in many parts of the country (like near concrete block factories!) and can be as strong and stable as poured concrete, *if installed correctly*.

Concrete blocks are available in various sizes and forms, but the most popular ones are 8", 10", and 12" wide, nominal. Nominal (in name only) means that they're actually a little smaller to allow for a mortar joint. An 8" × 8" × 16" concrete block is actually 7⅝" × 7⅝" × 15⅝". It makes it easier to figure out how many blocks you need.

Depending on the height, density, and load of the wall, concrete blocks may need reinforcement just as poured concrete does. Fortunately, concrete blocks have a couple of big holes in them so that, when stacked, rebar can be inserted and concrete poured into the holes to fill them. (Actually, rebar is vertically inserted into the footing before it cures, and the blocks are slipped over the rebar.)

Beginning at a corner of the footing, the concrete blocks are laid side to side with mortar below and on one side of the block. After a few rows are done and before the mortar dries, excess mortar is removed with a trowel, and a round metal tool called a striker is run along the joints to wipe away excess and smooth the joint.

If your house will have a brick veneer, the foundation will need a notch in the exterior that will serve as the footing for the brick.

Installing Piers

A pier foundation is more economical to build than a continuous wall foundation. However, it's typically only used for smaller houses in areas where pier foundations are allowed.

What's so simple about a pier foundation is that a hole is dug to the appropriate depth (below the local frost line), a cylindrical cardboard tube is inserted into the hole (and maybe some gravel in the bottom), and then concrete is poured into the tube to the plan-specified height. As it sets up, a J-bolt or other connector is inserted in the top of the pier and the whole thing is allowed to cure.

Alternatively, a stronger pier can be built with a large footing and pedestal as used in post-and-beam construction.

Pier construction.

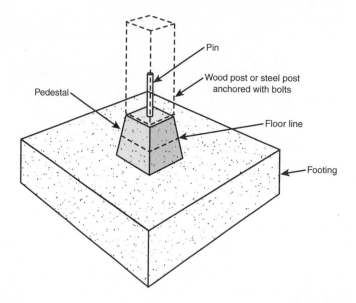

Piers are usually installed in rows below batter-board strings (see the earlier illustration in this chapter) that indicate the final height. Once in, beams are placed along the top of the piers, and the house is built from there.

Code Red

Make sure that your foundation's footing is below the local frost penetration line. It's just 1 inch in San Diego, California, and 60 inches (5 feet) in northern Montana, for example. Your local building code will specify the required foundation depth.

Other foundation materials include steel, wood, and stone. However, due to cost, life span, or load limitations, most local building codes don't allow them. Of course, make sure your lender allows concrete pier, wood, or other foundation materials before you go that route.

Making the Connection

You can also use pressure-treated lumber as the foundation wall if allowed by local building codes. The advantage is that it is easier for do-it-yourselfers to install. The disadvantage is that, typically, even pressure-treated lumber doesn't last as long as concrete.

If you're using the platform construction method, the pressure-treated wooden sill plate will be installed on top of the foundation wall (as you'll see in the next chapter). So anchor bolts, sometimes called J-bolts, must be inserted at the appropriate depth into the concrete or block wall before the concrete sets hard. Make sure there is sufficient exposed thread on the bolts so that you can later install the sill plate, washer, and nut.

Just to confuse matters, some construction plans call for steel straps to be imbedded in the foundation wall for attaching beams and joists. Do whatever the plans tell you to do.

If you're building a post-and-beam, log, manufactured, or other type of house, refer to the foundation plan and flooring plan for details on how the foundation walls are attached to the next level.

Installing Good Drainage

Water running down the side of a house can collect at the foundation and weaken it. So proper drainage is important to ensuring the life of your foundation.

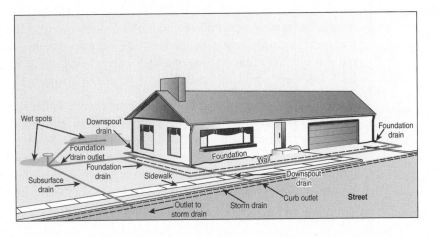

Drainage plan.

Once the foundation is cured and stable, coat the foundation with a sealant to keep moisture from seeping through to the inside. First, mark the grade level on the side of the foundation so that you know how far to treat the surface. Most sealants are tar-based. Ask your foundation or concrete contractor, or your materials supplier, for local recommendations.

Remember to treat the ground for pests while you can. This may mean treating the ground around the foundation against termites or other pests. We'll add termite shields to the subflooring in the next chapter.

Next, start replacing the soil up to the grade level. Based on the foundation plan, slope the grade away from the foundation and add gravel, pipes, drain tile, and culverts for proper drainage. Finally, finish grading away from the foundation for run-off.

Excavating for Septic Systems

Nearly one quarter of all homes use an on-site septic system to treat household wastes. The percentage is even higher among owner-built homes because many of them are rural or on acreage.

Household sewage is called black water, and drainage water (from sinks) is called gray water. They drain into a sealed septic tank that's located 2 feet below the ground surface. Overflow goes to the leach field for dispersion.

The septic tank size and leach field's capacity are based on the number of occupants the home will probably have (calculated on the square footage and number of bedrooms) and the quality of the soil. A two-bedroom house may have a 500-gallon tank, while a five-bedroom house may require a 1,250-gallon tank. Spend a few extra bucks and get an oversized tank and field.

Where does it all go? The tank's outlet is about an inch lower than the inlet so that the top scum and bacteria isn't removed. Sludge (don't ask) stays at the bottom. As the tank fills up, the excess goes out the outlet and along a bunch of perforated pipes to be disbursed by the leach field into the ground as ground water.

As you can see, septic systems need good planning. That's why, if your home will have a septic system (typically not allowed where there is a nearby sewer line), you'll need to get it planned and approved before you build. The building department will require a soil percolation, or perc, test to see how well it absorbs. From those plans, you, your excavator, and/or your septic system contractor will excavate for the system. Your local building department may also require that you identify an alternate leach field in your plans.

On the Level

If you need to hire a septic contractor, be sure to review Chapter 14. Because septic system contracting is a specialty, the contractor you select should be licensed for that specialty, depending on state regulations. Your grading, foundation, or plumbing contractor can recommend a septic contractor; in fact, they may be a licensed septic contractor as well as their primary specialty.

Excavation here means digging a hole big enough for the septic tank, and digging trenches to and from the tank. As mentioned in Chapter 17, don't install your septic system where heavy materials trucks may run over it and break it.

Alternatively, your local building code and soil analysis may require that you install a mound septic system or an incineration system. And they will tell you why. It's often because of a high water table on your property, because the soil isn't sufficiently porous, or just to be ornery.

Remember to dig trenches for draining away roof runoff from gutters and downspouts. Most use pipe to move the water away from the house, and then line the trench with gravel to dissipate the water. Some drain into a drywell or small underground tank for dispersion. You can also trench for irrigation lines now. Your landscape plan may include a lawn sprinkler system or drip line system. If so, you can start some of the trenching now, or hold off until you're ready to put in the landscaping.

Excavation and Foundation Resources

Valuable resources as you excavate for and install your home's foundation include the following:

- Local building department
- Building material suppliers
- Excavation contractors
- Transit-mix concrete suppliers
- Foundation contractors
- Backhoe and loader operators
- Construction engineers
- Concrete block manufacturers and suppliers
- Construction tool rental yards
- Alternative foundation sources (treated lumber, stone)
- The many websites, books, videos, and magazines listed in Appendix B

Your home's foundation is in. If you want to know even more about foundations, check bookstores or www.MulliganBooks.com for a copy of my book *Builder's Guide to Foundations and Floor Framing* (see Appendix B). It includes lots of illustrations, tables, and in-depth information on all types of residential and light commercial foundations. Next we'll look at how to install a subfloor system.

The Least You Need to Know

◆ Excavation is the removal of soil to make room for the building's foundation.

◆ Forms are containers into which concrete is poured for the foundation footing and walls.

◆ Concrete block walls can make foundation construction easier.

◆ Remember to treat the ground for pests, and install an adequate drainage system.

Chapter 19

Floors and Decks

In This Chapter

 ◆ Planning the floor framing
 ◆ Installing sills, girders, and joists
 ◆ Installing subflooring
 ◆ Planning for decks, garages, and outbuildings
 ◆ Finding resources for floors and decks

Your building lot is starting to take shape. From a weed patch, a foundation has sprung up. The next step is framing and finishing the floor.

This chapter takes you one major step toward finishing your home. It shows you how the floor framing is connected to the foundation, and then the subfloor is built on it. It also offers a look at how alternative (post-and-beam, timber, manufactured) homes construct the subfloor. It's more labor-intensive work, so you'll need some sports cream after reading this chapter.

Floor Materials

Most new homes built today use platform construction methods. That is, horizontal members called joists span the foundation walls, and then subflooring is installed over the frame to serve as a platform for building walls.

Here's how floors are framed. First, pressure-treated or redwood wooden sills are installed on the foundation perimeter with a seal and a termite shield below.

Next, because there's typically quite a distance between the exterior foundation walls, midfloor supports—usually concrete or block piers—are constructed with the foundation. Wooden timbers called girders are laid across these supports. Then wooden joists are installed perpendicular to the girders, spanning from one outside foundation wall to the other. They are nailed into headers or rim joists that sit atop the sills. Bridging or blocks are installed between the joists to keep them aligned. Finally, subflooring material (typically plywood) is laid on top of and fastened to the joists with glue and ring-shank nails.

Cross-section of foundation and floor framing.

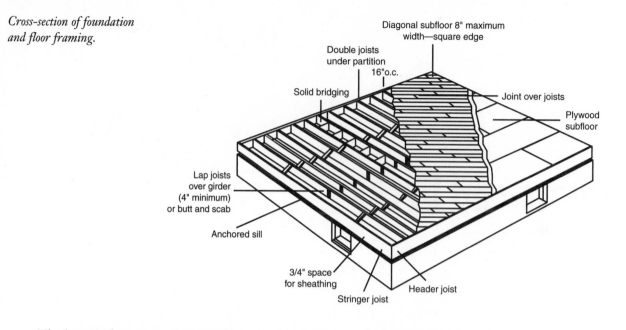

The house's floor is sometimes called its subfloor because a finished floor (carpet, hardwood, laminate) will eventually be installed on top of it. The framing contractor or carpenter usually builds the subfloor. That's because most of the job is cutting and fastening wood. Or you may do it. Or your GC. In any case, here are the typical steps for a platform home:

1. Verify that the foundation is level.
2. Install the seals and the sill.
3. Install the girders.
4. Install the joists.
5. Install bridging between the joists.
6. Install and trim the subflooring.

Of course, if your house is being built on a slab, it doesn't need a floor frame—unless it will have a second story.

If you're building a post-and-beam home or a log home kit, chances are the design calls for a platform construction floor, so this chapter will be useful.

Okay, that's an overview of floor framing and sheathing. Let's get more specific.

Installing Sills

Building Your Vocab

Sill, sill plate, and mud sill are all terms that refer to a piece of pressure-treated wood that's anchored to the foundation. It's the first component of the subfloor.

For most types of residential construction, the *sill* is the first wooden part of your house. Everything up to this point has been concrete and other materials. As the lowest wooden member, the sill is the most susceptible to moisture and termites. And, because everything above it sits on the sill, you don't want it to rot or be eaten by critters. So, your floor framing plans will probably call for a sill sealer on top of the foundation wall and below the sill. In most climates, it will probably require a metal termite shield as well. These two, and possibly other, components will need to be installed between the foundation and the *sill plate*.

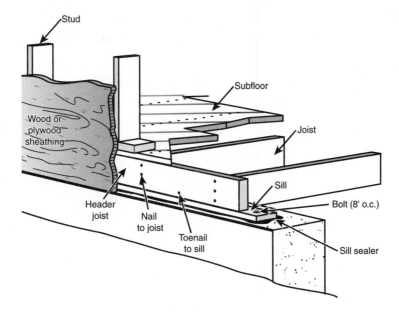

Components of a platform.

The sill or sill plate is attached to the foundation using anchor bolts you installed in the foundation as you were building it (see Chapter 18). These vertical bolts have an exposed thread. A corresponding hole is drilled into the sill plate, and then the plate is slipped over the bolt, and a washer and nut are installed. (Don't forget the sealer and shield!) You can also fasten the sill plate to the foundation using anchors for additional strength. A hole is drilled, and then a special anchor is inserted and tightened into the foundation. Alternatively, anchor straps or clips imbedded in the foundation concrete before it hardens are used to attach the sill.

CAUTION

Code Red

The sill plate anchor bolts and fasteners keep the house on its foundation. Make sure you install at least as many as are called for in your flooring plans.

Installing Underfloor Buildup

As I mentioned in the previous chapter, sometimes concrete or block foundation walls aren't built to a uniform height. This is especially true for houses built on moderate slopes. So how are these foundations brought up to level for the sill? Knee walls (also called pony walls) take up the slack. Why not just build the concrete foundation wall higher? Because wood is less expensive than concrete. And it can be just as sturdy.

Knee walls include the horizontal sill plate at the bottom and a top plate that's level with the highest foundation point. The vertical members between the sill and top plates are called cripple (meaning shorter than standard) studs.

For clarification, the sill plate is always the lowest part of the wooden frame. The top of a knee wall that's at the same level as the highest sill plate is called the top plate.

Installing Girders

Girders typically run the length of the building, offering support to the center of the floor. Depending on the depth of the house, you may use more than one girder.

The girder's job is to support one end of the joists. An outside foundation wall supports the other end. Your foundation and floor plans will indicate where and how the girder is to be installed.

How big should a girder be? It depends on the span or distance between supports. Use a 4" × 6" girder for a 6-foot to 8-foot span. Use a 4" × 8" girder for an 8- to 10-foot span. Use a 4" × 10" girder for a 10- to 12-foot span. Use a 4" × 12" girder for a 12- to 14-foot span.

Girders can be a single piece of wood, or a laminate of two to four pieces fastened together. You can also use steel girders. Girders typically sit atop a post that, in turn, sits on a concrete pier. The height of the girder member is usually the same height as the joists.

Fortunately, although they're heavy, girders aren't difficult to install. The ends sit in a notch built into the foundation wall or on a projecting post. Alternatively, fasteners are available to attach a girder to the foundation wall.

Girders in basements are typically supported by post jacks, supports that you can adjust to eliminate sag. You can also use metal girders. In fact, a manufactured home's underframe is made up of metal girders that will sit on top of the foundation walls or piers. Piers are then built under the girders for additional support.

Installing Joists and Bridging

It's coming together! Joist are 2" × 10" or 2" × 12" nominal wooden members, placed on edge, that span from a girder to a foundation wall. (Remember, nominal means they're actually a little smaller to allow for a mortar joint.) Spacing between joists is indicated on your floor plan, typically 16 to 24 inches depending on how much weight they must support.

Connecting joists to the rim joist.

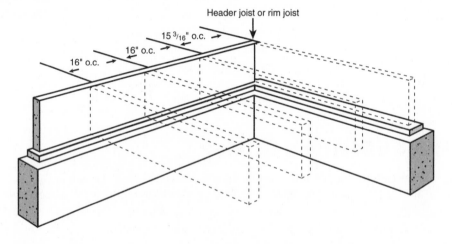

Before installing the joists, construct the band. A band (also called a box sill, depending on where you live) is simply the perimeter joists. Technically, the members that are perpendicular to joists are rim joists. The outside joists are also known as trimmers. Think of the band as the joists' outside frame. The band components are stood on their sides and nailed together to make a frame.

Then nail the end of each joist into the rim joist with three 16d nails. Follow the floor plan. Alternatively, use joist hangers, metal fasteners that attach the joist to the rim joist.

Some joists may have a slight bow in them, edgewise. Place them so that the crown (higher spot) is on top. A crowned joist will tend to straighten itself once the weight of the subfloor and floor is applied.

Your building plans may call for floor trusses, specially engineered components that are lighter in weight yet sturdier than conventional wooden joists. Or your plans may call for double-joists below load-bearing walls. That means you simply fasten two joists together so that they have double thickness and strength. If the plans call for plumbing under the wall, you may space the joists.

Make sure joist ends fall on and overlap at a girder. Then splice the joint with a piece of the same size wood or plywood, called a gusset. Or you can overlap the joists and nail them together.

How big should a joist be? It depends on the span or distance between supports. Use a 2" × 6" joist for a 6- to 8-foot span. Use a 2" × 8" joist for an 8- to 10-foot span. Use a 2" × 10" joist for a 10- to 12-foot span. Use a 2" × 12" joist for a 12- to 14-foot span. Make sure any changes comply with local building codes.

Some building codes call for bridging joists. That means components called bridges are installed to make a composite floor system. Solid bridging is installing solid pieces of wood, of the same dimensions as the joist, perpendicularly between the joists. Cross bridging is fastening smaller pieces of wood near the top of one joist and near the bottom of its neighbor. If cross-bridged, they are all fastened near the top first, after which the subfloor is installed, and then the bridges are fastened at the bottom.

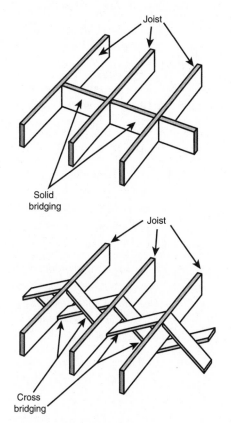

Solid and cross bridging.

Cross bridging can also be metal fasteners specially built for the job. Check your building plans. Typically, one end is nailed to the top edge of one joist and the other end to the bottom edge of the neighboring joist.

In some cases, you'll want an opening in the joists. For example, a stairway to a basement will require an opening in the floor for the stairs. The opening requires a double header at both ends of the opening. A double header is simply two pieces of wood that serve as a header into which the joists are nailed at the opening.

Joists can be extended outside of the foundation to become the frame for a deck. These are especially popular for balconies. Make sure there is sufficient drainage from the flat deck or balcony.

Special joists are needed for overhangs, sunken floors, and other designs. For example, a bay window may protrude from one wall in your design. Because there is little weight or load to it, it doesn't need a foundation under it. So floor joists are extended outside of the foundation wall, and are framed to support the window structure. It's called an overhang or a cantilever.

The joist plan may also include modifications for special plumbing, the HVAC ducting, or other features. Your flooring plans will offer the specifics. Some types of alternative construction, such as log homes, use large girders instead of joists. The same general rules apply.

Installing Subflooring

Before installing subflooring, make sure any work that needs to be done underneath the floor has been or can be done. The plumber needs to install pipes. The HVAC contractor may need access for vents. Is it easier to install insulation now? Check first.

Subflooring is installed over the floor joists as a working platform and a base for the finish flooring. Subfloor can be 1" × 6" or 1" × 8" tongue-and-groove boards or, more popular, ½" or ¾" plywood.

Subfloor plywood is selected based on the load it will support and whether it will be subjected to water. By using sturdier floor sheathing, you can space joists farther apart. For example, 48" o.c. (on center, or from center to center) joists are used with special 1⅛" plywood. You may be able to save some money.

Some species and thicknesses of plywood can offer support above 24" o.c. joists. Your plans may call for exterior grade (with waterproof glue) for bathrooms and the kitchen, or all through the house.

Install plywood with the grain direction of the outer plies at right angles to the joists. Stagger the plywood so that end joints in adjacent panels break over different joists.

On the Level

If you're having trouble keeping track of where to nail the flooring to hit the joists, use a chalk-line string (available at hardware stores) to mark the flooring, attaching one end to a nail directly above the end of a joist. Directions for use typically are on the carton.

Floor sheathing is usually started at one corner of the floor band and installed perpendicular to the joists. The next course of sheathing begins with a half sheet to stagger the joints. Use subfloor glue and ring-shank or screw nails for a silent floor. (You can save time by renting a compressed-air or electric nailing gun to apply the subfloor.)

Your building plans will indicate nailing spacing. Typically, it requires 8d common nails at 6-inch intervals into the edge joists and 10-inch intervals over other joists. Nailing patterns may vary depending on joist spacing and on whether you will be adding an underlayment prior to installing the finish flooring.

Finally, you may need to notch or cut holes in the subfloor for pipes, drains, and underfloor utilities.

Second Floor

Framing the second floor will be almost identical to framing the first floor. Of course, you'll need the walls framed first! That's covered in Chapter 20.

Once you frame the first floor walls, you build a band, and attach joists, just like on floor one. You add bridging, if required. Then you install the subflooring. Special framing is required for stair openings and any fireplace chimneys.

Framing Decks, Garages, and Outbuildings

If your plans call for a deck, you can construct it now or later. If now, you can build it by extending the joists from the house in a cantilever design. This method is popular for elevated decks or balconies. The building plans will give you the details. Or you can build a separate frame and attach it to the foundation or to the sill band. This method is used for both new construction and retrofits (adding it later).

Building plans for garages and other structures may be similar to those for your residence. For example, an attached garage may have a concrete slab floor with a short continuous-wall foundation. Or it may be framed just like your house except that the joists may be spaced closer to support the additional weight of vehicles.

Outbuildings may have slab floors as well, or may have joist floors with wider spacing depending on the structure's load. In any case, follow your building plans and the instructions here and in Chapter 18 to get the job done.

Flooring and Deck Resources

Need to know more about installing the subfloor? Conventional and post-and-beam kit homes typically don't include the foundation or subfloor. However, they usually include specific plans for its construction.

Additionally, *Carpentry & Construction*, *Third Edition*, by Miller, Miller, and Baker is highly recommended (see Appendix B). At nearly 700 pages, it includes a more detailed description of the construction process.

You can also get help from the following:

♦ Make sure your building plans include extensive detail so that you and/or your crew know exactly how to frame the floor.

♦ Hire a local framing contractor to advise you as questions arise.

♦ Hire a retired carpenter to mentor or help you to frame your house's floor.

♦ Contact local floor truss suppliers to find out more about their products and whether they can make construction easier.

Chapter 20 will show you how a house quickly takes shape with framing.

The Least You Need to Know

♦ Platform construction floor framing is used for conventional homes, kit homes, log homes, and post-and-beam homes.

♦ The sill or mud sill attaches to the foundation and supports the walls.

♦ Girders support the center of the floor for the length of the house.

♦ Joists are installed at right angles to the girders and support the floor.

♦ The subfloor is the flat surface on which you build walls and other components.

♦ Decks and outbuilding floors are constructed similar to residential subfloors.

Taking Shape: Framing

In This Chapter

- ◆ Selecting framing materials
- ◆ Framing walls, doors, and windows
- ◆ Framing for fireplaces, skylights, and other special features
- ◆ Framing roofs, stairs, garages, and outbuildings
- ◆ Finding resources for framing your home

A frame defines and encloses an object: in this case, your house. Framing is the job of building that frame. Makes sense.

In fact, framing *does* make sense. It's a very logical task in building your own home—once you understand how it works.

That's what this chapter is all about: making framing logical. It will show you the steps to framing walls, doors, windows, ceilings, roofs, skylights, and other components. It will also show you how to select materials.

Fortunately, your house plans will include a framing plan and even a wall section to make the job easier. So strap on your tool belt and let's start having some fun.

Framing Materials

The primary ingredient in framing most homes is lumber. Lumber is dimensional wood cut to a specified size to make construction easier. Most lumber is cut to 8', 10', 12', or 16' in length, with one big exception. Studs are 2" × 4" nominal. But studs are also shorter than standard, typically 92¼" (7'8¼").

On the Level

Remember: Actual dimensions (width and depth) of lumber is less than the nominal dimensions, depending on the type of wood and whether it is dried or not. For example, a 2" × 4" nominal may actually measure 1⅝" × 3⅝".

Why? Because the height of the sole plate, stud, top plate, and double-top or cap plate together is 8'¼". That gives you room to install the ceiling and wall drywall panels without cutting them. This is called production framing.

Your house's materials list will indicate how many studs and boards are needed for the job. Of course, it's an approximation. And you may decide to not use some of the delivered lumber depending on its condition.

Lumber grades are important, too. The materials list will probably indicate which grades you should be using. Much depends on where you live and what species and grades are most popular and available in your building area. Douglas firs of the west will be graded differently than southern pine or northern firs.

Lumber will be delivered to your building site in a bundle or on a pallet, tied for easier transport. Make sure you cover it in inclement weather, and keep it bundled until you need it, to deter theft. Kiln-dried lumber is more expensive but has less moisture than green lumber.

Depending on your materials supplier, you can return unusable framing lumber to the supplier for credit or replacement. That means checking each board before you use it and setting any unusable boards aside. You may be able to get shorter boards, such as those for under window frames, from the discard pile. Some lumberyards will charge a restocking fee to replace the boards.

If you've hired a framing contractor, coordinate delivery so that everything is there when needed. Time is money. Framing contractors will bid based on square footage of the framed area, plus extra fees for special jobs like framing stairs.

Ka-ching!

Because framing requires as much manpower as skill, many owners hire themselves to their framing contractor to help out and save money. A typical framing crew is two or three people including the lead framer or boss. If you're hired as a helper, be prepared to work harder than more experienced workers.

Lumber isn't the only residential framing material. Post-and-beam framing is also popular and will be covered later in this chapter. In addition, steel framing is increasingly popular, but because of proprietary systems, installation steps are outside of the scope of this book. Fortunately, steel framing is similar to lumber framing, and suppliers can offer you specific instructions for installation.

Log homes use horizontal logs stacked on top of each other for exterior walls. The log home system you purchase will include instructions for joining them. (My out-of-print book *Building a Log Home from Scratch or Kit, Second Edition*, includes full instructions.) Most interior walls in log homes are built using conventional lumber and methods so that wall plumbing and wiring can be included.

Framing Walls

Framing your house is a simple process:

1. Build the longest exterior wall on the platform.
2. Erect the wall and brace it.
3. Build the exterior wall facing the longest wall.
4. Erect that wall and brace it.
5. Build and erect the other exterior walls, joining them at corners.
6. Build, erect, and attach each of the interior walls according to the plans.
7. Frame the roof, including the ceiling.

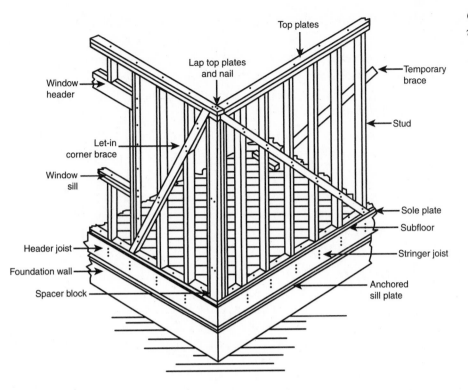

Components of a platform wall.

Top plates

Lap top plates and nail

Window header

Temporary brace

Let-in corner brace

Stud

Window sill

Sole plate

Subfloor

Header joist

Stringer joist

Foundation wall

Spacer block

Anchored sill plate

If you're adding a second story, build a platform floor above the first floor's walls, and then build walls above it before framing the roof. Your house's framing will offer the specific details.

Will you install a large one-piece shower or other larger-than-a-doorway components in your house later? If so, save yourself some frustration and place it on the floor platform after exterior walls are framed and before they are all erected.

That's about it. Of course, the dirty work is in the details. Fortunately, this chapter will describe each of the steps.

Before framing your first wall, take a look at the framing plans. Notice that most of a wall is made up of vertical studs standing between horizontal lumber called *plates*. The exceptions are openings for windows (with shorter studs) and places where two or more studs are nailed together. The multiple studs are used to reinforce corners and where an interior wall joins an exterior wall. Plates and studs are typically nailed together with 16d nails.

Building Your Vocab

Plates are horizontal framing members. A sole plate is the bottom horizontal member of a wall. A top plate is the top horizontal member of a wall.

So, here's what professional framers do: They lay the two plates together on the platform and, following the framing plan, mark *both* plates at once for the location of studs. Once marked, they separate the two plates on the floor platform and place the studs at right angles in between the plates. Next, they nail through the plates and into the studs. Make sure that the plates always end in the middle of a stud, even if you have to trim a plate to make it happen. Stagger the joints between the top plate and the double-plate or cap plate.

The ends of walls that become the corners use a little different structure, typically two full studs with short spacers in between to give them the width of three studs for support.

After all the nailing is done, the frames are lifted into place to become a wall frame. They are temporarily braced so that they don't fall down before other exterior walls are attached. The most popular method to temporarily brace a frame is to nail one end of a 12-foot board to the side of the erected wall near the top, and then nail the other end to the top or side of the platform. Once two walls are joined, corner braces are installed on the exteriors. Plywood sheathing can also be nailed to the outside as bracing.

Don't forget to use a carpenter's (long) level to make sure the walls are plumb (perpendicular) before permanently nailing them in place.

Interior walls are framed and installed in much the same manner as exterior walls. One major difference is how the walls are attached to each other. The top plate butts the perpendicular wall, and the cap plate overlaps and is nailed to it.

That's it, in a nutshell. All exterior and interior walls are built in the same manner, first horizontally on the floor, and then one side is lifted to become a vertical wall.

Fortunately, your framing plan shows you exactly where each of the studs is to be located, where the door and window openings are, and where walls join each other.

Framing Doors and Windows

Framing openings for windows and doors is a slightly different procedure. A door opening, as you can imagine, doesn't have studs in it. So what supports the weight of the roof? A header is installed above the door opening where extra studs (called trimmers) support it. The header is typically larger dimensional lumber and framing in order to place it at the correct height. Some building plans allow the header to be replaced with short studs called cripples. Check the plans.

Framing for a window.

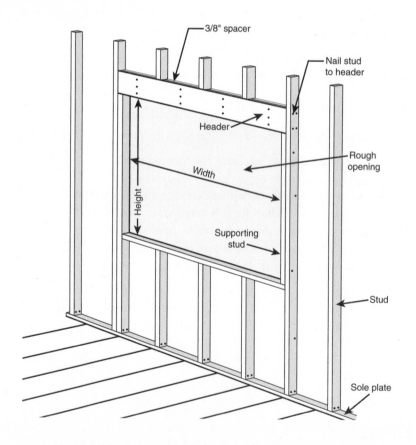

Header lumber must meet the local building code. In most cases, that means using 4" × 6" for openings up to 3'6", and 4" × 8" for up to 6' openings in one-story houses. Two-story construction will require larger material.

A window opening, too, has a header. But because the bottom of the window is above floor level, it also needs support below the window. The support is a horizontal double-stud called a sill, supported by short studs or cripples.

Standard stud spacing is 16" o.c. (on center, or between centers). However, 24" o.c. is allowed under some building codes if the sheathing (wall covering) takes up some of the load or if the walls are thicker (2" × 6"). Your framing plans will give the specific measurements for framing walls, doors, and windows.

Most doors are 6'8" tall. The actual opening will be 2 inches taller depending on the trim and flooring used. The rough opening for windows is typically the glass width plus 6 inches, and the glass height plus 10 inches. Plan details for your house will include specific measurements.

Once the walls are built and in place, cut and remove the sole plate at the bottom of door openings.

> **On the Level**
>
> Your window schedule may include a "2442" or other listing. It means that the window sash is 2'4" wide and 4'2" high. The window opening in the wall frame will be larger depending on whether the window is wood or metal. Check the plans or ask your window supplier about rough opening dimensions.

Most houses are built using what's called light frame construction methods. If your house uses 2" × 6" or thicker walls or carries heavier-than-normal loads from a second or even third floor, or from a heavy roof, the framing rules may change. Your building plans will then include a variation on the construction methods described here. (Remember that 2" × 6" walls may require special doors and windows that fit into the deeper openings.) However, the process will be about the same.

If you're building a pre-fabricated house, the walls will come in prebuilt sections, framed and ready to go. So all you'll need to do is raise and attach the walls according to the manufacturer's instructions.

Special Framing Considerations

Your unique home design will probably have some amenities that require special framing. For example, fireplaces, skylights, bay windows, sunrooms, and other features need framing. Fortunately, what you've learned thus far about traditional house framing will serve you well.

- Fireplaces are typically framed as an opening just like a big door with a header and trimmers. Some will require cripples for the headers. Fireplaces may also require framing an opening in a wall or ceiling and roof for the chimney.
- Skylights are simply windows in the roof. Framing them is similar to framing wall windows.
- Bay windows, bow windows, and other windowed appendages are framed like a large doorway with windows in it.
- Sunrooms are small rooms that protrude through a wall to the exterior. They may be framed as the walls are, or the opening may be covered until the sunroom is installed later in construction.

As with other variations, refer to your framing and building plans for specific dimensions, components, and construction detail.

Framing Roofs

Unless you're building a second story, framing the roof is next. Roof framing is similar to wall framing. In fact, many of the terms—joists, cripples—are the same.

Roof framing components.

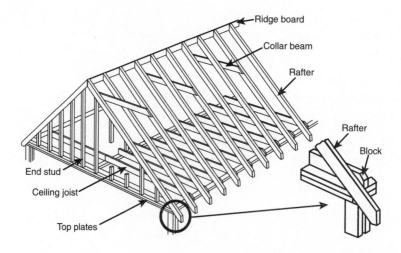

As you've learned, a joist is the horizontal bottom of the roof frame. The diagonal components angled to the roof pitch are the rafters. The short studs between the joists and rafters are short, so they are called cripple studs or cripples. Finally, at the top of the roof, the ridge board connects the roof frames together.

You attach the various roof frame components to one another using special metal or wood plates called gussets, or you can use glues.

As you build a roof, you'll hear four common terms:

◆ The **pitch** is the slope of a roof from the ridge to the top plate, typically expressed as a ratio of the rise to the span.

◆ The **rise** is the distance from the bottom to the top of a roof frame.

◆ The **span** is the distance between a building's outside edges that a roof covers.

◆ The **run** is the horizontal distance between the roof peak and one outside edge that a roof covers.

Depending on the complexity of your home's design, you may be able to build the roof frames on the platform and erect them atop the walls. A pattern, or jig, is marked out on the platform to standardize the frames and make them easier to build. Additional frames are made following the pattern piece.

More complex roofs may require engineering, depending on local building code. In that case, special roof frames or trusses are built in a factory and brought to the building site. Your approved building plans will indicate whether the roof will be framed on site or built off site.

Trusses are relatively easy to install. They are hoisted to the top of the framed wall, and then moved into position and installed by hand. Installation typically means making sure the truss is correctly placed and that there is the designated space between them. Trusses are attached to the top plate of the wall using a metal bracket and/or nails.

Kit homes typically include pre-built trusses. So do some log home kits; however, the materials may be logs instead of dimensional lumber. Post-and-beam homes may use engineered trusses or site-built roof frames.

Dormers, a protrusion from the roof, require special framing so that they safely tie in with the roof (see the following illustration). The building details sheet will include the specifics.

In drier regions, flat roofs are allowed. Construction is relatively simple, following the construction of the floor (see Chapter 19).

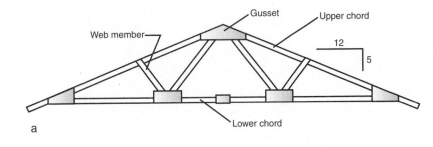

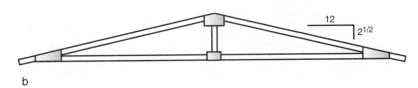

Common wood truss designs:
a) W-type, b) king post,
c) scissors.

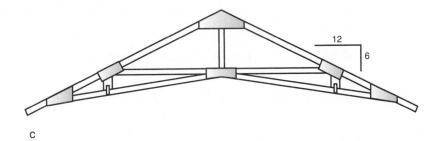

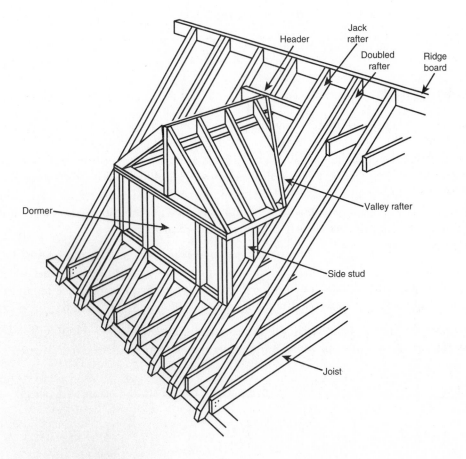

Components to dormer
framing.

Framing Stairs

The first step to installing stairs is to make an opening in the floor, called a stairwell. It may be a hole in the first floor for stairs to and from a basement, or it could be for stairs between the first and second floors.

The stairwell opening is framed by installing two cross-members (headers) between two joists. The opening should be slightly larger than needed because it's easier to make it smaller later than to enlarge it.

If your plans call for stairs, dimensions will be included, making the job easier. The place where your feet land is called the tread. The part that your toes might hit as you go upstairs is called the riser. The side support for the stairs that holds the treads and risers in place is the stringer. If the stairs turn, plans will call for a flat area called a landing.

Most standard homes have a straight stairway comprised of 14 steps, including a tread or run of $10\frac{3}{8}$" and a rise of $7\frac{1}{2}$". The tread run may vary depending on the house plan, but the $7\frac{1}{2}$" rise is common. Typical stair width is 36" with 32" a minimum.

Ka-ching!

Framing stairs can be a challenge, even for professional carpenters. Fortunately, layout tools such as framing squares are available to help them—or you—make the calculations needed to do the job right the first time.

If you're carpeting the stairs, all you'll need is a sturdy stair frame over which you will apply the carpeting. If you plan to finish the stairs with wood and fancy rails, you need to give the components a base on which to build (as detailed in Chapter 23). Here are the steps to building the typical stairs:

1. Cut two stringers (also known as carriages) per the plans.
2. Attach the stringers solidly in place.
3. Install the treads.
4. Install the risers (if needed).

Once installed, the stairs will later be painted, carpeted, or finished in decorative wood.

Post-and-Beam Framing

Post-and-beam construction uses posts (4+") rather than dimensional lumber (2+"), spacing them farther apart. The horizontal components are the beams. Instead of 16" to 24" spacing, posts are installed 4 to 12 feet apart because these larger components can handle more load. And therein lies the main advantage of post-and-beam construction: more space between vertical supports. This allows for larger windows and skylights. In addition, better-quality wood is often used and is exposed to give the house a more natural look than conventional homes. Because of their bulk, post-and-beam homes don't use traditional roof and rafter systems. They don't need to, because so many of them offer exposed roof members.

Posts and beams are attached to each other in various ways. One popular method is called mortise and tenon. The mortise is a shaped hole into which the tenon is inserted, locking the two components together. The joint is fastened with glue or nails. Alternatively, the posts and beams are attached to each other and to the floor using brackets and bolts or screws. Mortise-and-tenon homes typically use 8" × 8" members.

Post-and-beam frame homes are finished by filling in the space between posts with more traditionally built walls. Because they are often hung from the horizontal beams, they are sometimes referred to as curtain walls.

That's an overview of post-and-beam framing. If you're buying a system or a kit, the manufacturer will provide specific instructions for framing.

Framing Garages and Outbuildings

Garages and other outbuildings are framed similar to residential structures. The main difference may be the spacing of wall studs. Unoccupied buildings don't always require the same structural support as those we live in. However, you may opt to build an attached garage or other building to residential code in case it is some-day remodeled and occupied.

Decks are framed in one of two ways. A cantilever (see Chapter 19) is an extension of the floor joists beyond the foundation wall. This cantilever can be the basis for a deck or balcony, requiring only the installation of posts and rails to finish it off. Or a deck can be built and attached to the side of the house at a doorway, and then finished with decking, rails, and stairs.

Framing Resources

Framing a house is a learnable craft. If you've decided to do your own framing, take a local class, read a book, or volunteer for a Habitat for Humanity project. Many community colleges have hands-on classes on residential house construction and framing. *Carpentry & Construction*, *Third Edition*, by Miller, Miller, and Baker (see Appendix B) is an excellent reference. You can find your local chapter of Habitat for Humanity International online at habitat.org or by calling 229-924-6935.

Next, we move to the exterior of your house, getting closer to moving in!

The Least You Need to Know

- ◆ Many owner-builders save money by doing some or all of their home's framing.
- ◆ The framing plan offers specific details and dimensions for framing your home.
- ◆ Plan openings for doors and windows in advance to save time and money.
- ◆ Fireplaces, skylights, sunrooms, and stairs require special framing.
- ◆ You can frame roofs on site, or build with fully assembled trusses.

Chapter 21

Closure Is Good: Exterior

In This Chapter

- Selecting exterior materials
- Installing sheathing and siding
- All about roofing and gutters
- Painting the exterior
- Finding resources for exterior work

You've successfully framed your house, so now it looks like a wooden skeleton. It needs skin! A house's "skin" is exterior sheathing. It includes a vapor barrier, siding, roofing, and even paint (call it "makeup"). This chapter includes these and related topics.

By the time you're done with this chapter your house will be ready for you to start working on the inside, turning it into a home. Let's go to work.

Exterior Materials

Exterior materials include vapor barrier, wall sheathing, siding, roof sheathing, roofing, soffits, and trim. Your house's materials list will usually include a comprehensive list of what's needed. If not, ask your building materials supplier or advising contractor to make up the list using the building plans.

Vapor barrier is a thin film that protects the walls from moisture. It can be special plastic sheeting, reflective foil panels, or builder's felt (tar paper). Which one your house needs depends on local climate and building codes. Homes in northern Minnesota will require a more efficient barrier than houses in Southern California.

Siding comes next. Here's where there's lots of variety. Siding can be 4' × 8' textured exterior plywood panels, horizontal lap siding, brick, stone, stucco, adobe, or a log facing. Some horizontal siding is made of solid wood or wood products, others of vinyl or even concrete.

Roofs are typically sheathed (covered) using exterior plywood panels. Builder's felt is then applied. Finally, the roofing material is installed over it. The most common roofing material is asphalt shingles. Others include wooden shingles, shakes (handmade shingles), and masonry tiles. Additional roofing materials can be used depending on local building codes.

Soffits are the underside of the roof's overhang. Depending on your home's design, they may be open (log and timber homes) or closed (conventional homes). Closed soffits use plywood to cover the end of the roof rafters. Vents are typically installed to allow warm air from under the roof to escape. Your home's location and construction may vary.

The exterior is painted after all the doors and windows are installed. Finally, trim (called fascia) is often applied around the exterior of the roof to enclose the soffits and to hang the gutters.

Before I get into the specifics of exterior finishing, let's discuss safety. Exterior work can be some of the most dangerous on the building site. Why? Because of poor ladders and scaffolds. Fortunately, most ladders and scaffolds you buy or rent include instructions for safe operation. However, many tradespeople—and owner-builders—ignore the rules and do some really dumb things. So here are some basic safety rules for using ladders and scaffolds:

- Make sure the ladder or scaffold your buy or rent is safety rated for your job (check the sticker on the unit for specifications and limits).
- Inspect the ladder or scaffold for condition before using it.
- Make sure the ladder or scaffold is clean with no grease, wet paint, or other materials on it.
- Set the legs on a firm and level surface.
- Don't lean ladders sideways.

Code Red

NEVER set ladders or scaffolds up near power lines!

- Make sure the ladder is one fourth as far from the wall as it is tall (for example, 3 feet away for a 12-foot ladder).
- Make sure that scaffold platform boards are solid and will hold the weight of people and equipment on it.
- Don't try moving a ladder or scaffold with materials or tools on it.

Installing Wood Siding

Vapor barriers are relatively easy to install. Rolls are typically installed horizontally around the perimeter of the house, nailing or stapling to the wall framing according to the manufacturer's directions. Solid sheet vapor barriers including reflective foil sheets are installed as panels nailed or stapled to the wall framing. Builder's felt is attached in rows, beginning at the bottom of the wall with each successive row overlapping the previous. They, too, are nailed or stapled to the wall framing.

On the Level

Some construction methods install the doors and windows (aluminum and vinyl) before siding is installed, while others don't. And some builders prefer to side *and* paint the house before installing windows and doors. Base your order on what makes the most sense for your home style and building materials, as well as on the preference and availability of materials and trades people.

Sheathing comes next. Depending on your plans and local building code, this is typically 4' × 8' exterior-grade plywood sheets. Doors and windows aren't covered, but just about everything else is. In some climates, sheathing is not required, so siding is applied directly over the vapor barrier. Or the vapor barrier and sheathing are included in a single product.

As I mentioned, many types of siding are used on modern houses, including wood siding, textured panels, fiber cement, vinyl, aluminum, and bricks. Installation depends more on the size of the product rather than on its materials. For example, horizontal siding is installed starting at the bottom of the wall, and overlapping each course, going up the wall. Panel siding is typically installed in 4' × 8' or larger sheets, interlocking one to the next. Doorways and window openings are left uncovered for obvious reasons.

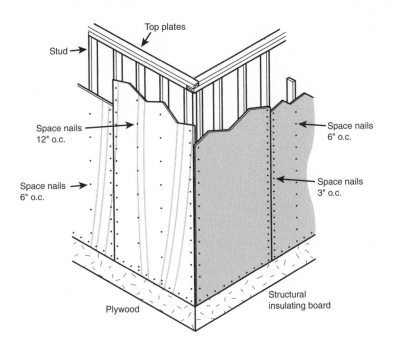

Installing plywood or insulation board sheathing.

How much horizontal siding will you need? When calculating, allow for the overlap (typically 1½") and make sure you use the siding's actual (¾" × 5½") rather than nominal (1" × 6") dimensions to figure the exposed siding (5½" minus 1½" equals 4"). It's easiest to calculate the entire wall, and then deduct for windows and doors.

How much panel siding will you need? That's much easier. It will take 10 4' × 8' panels to cover an 8' × 40' wall (320 s.f.). Windows and doors will be cut out of the panels.

How you install the sheathing and siding can be very important. For example, your approved building permit and plans may require a specific nailing or fastening pattern. Make sure you know what the plan and building inspector requires.

Installing Masonry Veneer

Veneer is a covering. Masonry is mineral-based building materials. So masonry veneer is a mineral-based wall covering. Prime examples are natural stone, clay brick, concrete block, and stucco or adobe.

Masonry veneers are typically installed over plywood sheathing with a vapor barrier. Because masonry products are heavier than wood, a solid footing is required for the materials. In most cases, the foundation is built with an exterior lip that serves as a footing for the stone, brick, or block. The footing's width depends on the materials that will be used. Fortunately, your building plans include details on how the masonry veneer is installed.

Clay bricks are usually 4" × 8" × 2¼" nominal (3⅝" × 7⅝" × 2¼" actual), surrounded by a ⅜" layer of mortar. The running bond pattern that overlays each row by one-half brick is the most common.

When building a brick veneer, follow the details to make sure weep holes (unmortared side joints) are included to allow for drainage. Also add a corrugated fastener every few rows to attach the brick mortar to the sheathing for stability.

Ka-ching!

Do you like stonework, but don't want to go to the hassle or expense of covering the entire exterior with masonry? Use it sparingly. Our house uses stonework around a front bay window to accent the home's redwood-shingle siding.

Concrete blocks are commonly 8" × 8" × 16" nominal (7⅝" × 7⅝" × 15⅝" actual), surrounded by a ⅜" mortar joint. Mortar is laid on top of the footing, and then the corner block is installed, followed by mortar and another block. Each course gets a layer of mortar in short sections so that it doesn't dry.

Stone veneers are built in a similar manner except that stone is irregular and requires some skill to place the individual stones. If you're doing it yourself, practice where it won't show, doing the front of the house last.

Stone mortar is similar to that for block and brick. You can purchase it in a premix bag, and just add water. You can mix it by hand or use a rented mixer.

The trick to building stone walls is to "face" the front, or keep the front of the wall straight. Fill the back side as needed with smaller stones and mortar for stability.

Or you can cover the exterior with mortar. Then it's called stucco. First, fasten metal lath or screening to the sheathing. Then mix and use a trowel to spread the mortar on the rough surface. For best drying, apply it in layers, smoothing out the final layer. Don't mix pigment into the stucco, because the color will probably come out uneven. Instead, use natural mortar, and then paint it when done. Professional stucco contractors apply a ½" to ¾" scratch coat, and then a brown coat to smooth the wall, followed by a color coat, allowing adequate time for each coat to dry.

Many homes that look like they are built of adobe are actually contoured stucco. In fact, adobe is simply a clay mortar spread over a frame of clay blocks.

Roofing

Time to roof your house! Refer to the roofing or building plan for specific materials and dimensions. The building details sheet(s) will probably show you roofing details such as underlayment, materials, and nailing pattern.

Roof sheathing is similar to wall sheathing. Most homes use exterior plywood sheets to cover the rafters and give a solid base for the roofing materials. Some types of tile and shingle roofs use horizontal wooden strips nailed to rafters, with space between each strip for air circulation. Check the approved building plans.

There are many popular styles of roofs including flat, shed, gable, hip, gambrel, mansard, and combinations of these. Most are easy to identify. If the roof has two sides, it's probably a gable roof. If it has four, chances are that it's a hip roof. If it has four and the top is flat, then it's a mansard roof. If there's a smaller roof (valley) at right angle to the main roof, that's either a gable-and-valley or a hip-and-valley roof.

Besides aesthetics, what's the purpose of a roof? It keeps water from puddling on top of the house. In dry regions, that's not a problem, so flat roofs are okay. In other areas, local climate tries to dictate the preferred roof styles.

So drainage is an important issue when designing and building a roof. That's why special metal or plastic components called flashings are used to make sure water doesn't seep into the house. Flashings that are located where rooflines join are called valleys. Other flashings are installed around chimneys, vent pipes, and skylights. Especially skylights!

Ka-ching!

Professional roofers install both flash and counter flash. Flash is L-shaped and fits in the valley between two roofs or around a chimney. Counter flash is mounted at the top edge of the flash to keep water from seeping behind it. Cheap insurance!

Roof pitch is important, too. The pitch of a framed roof (see Chapter 20) is designed to make sure that water, snow, or other stuff doesn't stay on the roof longer than necessary.

Roofing materials are typically installed beginning at the lower edge of and across the roof. The materials can be fiberglass or asphalt shingles, wooden shingles, shakes, tile, or other materials. Each course is installed to overlap the top of the lower course. The roofing is also staggered to keep moisture from getting under the material and into the house.

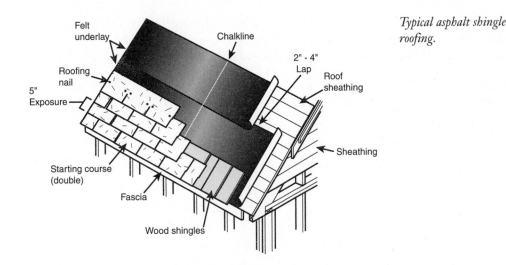

Typical asphalt shingle roofing.

Flat roofs, of course, are roofed a little differently. Most roofers overlap builder's felt, and then cover it all with hot asphalt or tar.

How much roofing material will you need? It's relatively easy to figure for simple flat or gable roofs. It's a little more complex for gable- or hip-and-valley roofs with so many angles. If your materials list isn't any help and you really need to know, your building materials supplier can help you calculate the needed materials.

On the Level

Roofing materials are commonly sold in "squares." A square of roofing is the amount required to cover 100 square feet. (That's three bundles of standard fiberglass shingles and one roll of 90-pound mineral-rolled roofing.) Get the total square feet of roof surface and divide it by 100 to get the number of squares needed. For complicated roofs, add 5 to 10 percent for waste. For example, if the roof surface is 2,000 s.f., you'll need 20 squares of shingles and 20 rolls of roofing paper. Add one or two squares and rolls for complex roofs.

Cornice and Soffits

The cornice is the area between the exterior wall and the edge of the roof. It includes the soffit, or underside of the roof, and the fascia board, or edge. The cornice is also called the eave, or overhang, of the roof.

The two most common types of cornices are open and closed or box. Most conventional designs use box cornices. Post-and-beam, log, and many other designs with larger structural members show them off with open cornices.

Open cornices use blocks to close off access to the area under the roof. Some blocks are replaced with vents for air circulation. Closed cornices typically include vents to allow air trapped below the roofline to ventilate. The vents can be single or continuous.

Your building plans will indicate the type of cornice, materials, dimensions, and other details. If not, you'll need some professional help to design and possibly to install the cornices.

Besides traditional plywood soffits, aluminum soffits are becoming more popular. One version uses 50-foot rolls of aluminum soffit material, available in widths from 12 inches to 48 inches. Tracks are installed on the cornice, and the material is inserted from one end of the soffit toward the other.

Some plans include a fascia or facing board, and others omit it. They are used as trim and as partial support for gutters. Additionally, some designs will include a metal roof drip edge that keeps roof runoff from slipping below the first course of shingles and into the house. The drip edge is installed over or under the top edge of fascia board. Some designs also add decorative trim, called gingerbread, around the edge of the roof, either attached to or replacing the fascia board.

Installing Gutters

Gutters collect runoff from the roof and channel it into downspouts that direct it away from the house. Gutters and downspouts are available at larger building materials retailers, as well as by hiring a specialist. Gutters used to be built as troughs made of wood. Today's gutters are typically of aluminum with baked-enamel finish or vinyl.

Gutter and downspout system components.

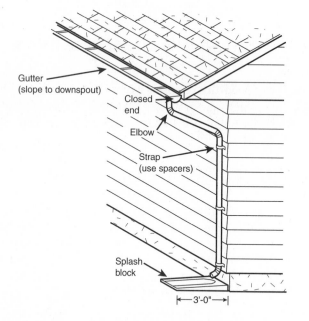

Most do-it-yourselfers can install a gutter system for the typical house in a weekend, so this may be done while the home is being built or once it is finished. One question is whether the building inspector will issue an occupancy permit if the gutters aren't installed.

Here are the steps to installing the typical gutter system:

1. Measure and purchase the gutters, end caps, hangers, downspouts, leaders, splash blocks, and fasteners needed for the job.

2. Mark the fascia board (or rafter ends) to indicate the top edge of the gutter, sloping the line slightly (1 inch per 20 feet is typical) toward the downspout.

3. Install the gutter hangers according to the manufacturer's instructions, usually every 2 to 4 feet.

4. Measure, cut, and assemble the gutter, and install the downspout mouth.

5. Install the gutter run on the hangers.

6. Install the downspouts.

7. Install leaders from the base of the downspout to a drywell (a 4-foot hole filled with coarse gravel) located at least 6 feet from the house's foundation. (Some locations will require that runoff go to a ground sewer.)

8. Test the system with a garden hose and seal any leaks.

Alternatively, a gutter contractor can install the system for you. Most have the equipment on the back of a truck to fabricate seamless gutters in your driveway and have the tools to install them.

Painting Your Home's Exterior

A home's exterior may be painted prior to or after the doors and windows are installed (see Chapter 23). However, we'll cover the topic here as it relates specifically to the exterior of your home.

Painting the exterior can save you money. It can also be a cause of frustration if you don't have the right tools and skills. In either case, knowing how home exteriors are painted will make you a smarter consumer. (Painting home interiors will also be covered in Chapter 23.)

Three types of paints are commonly used for house exteriors. Which one your home gets depends on the siding, local climate, and who's doing the job. Some are easier for do-it-yourselfers than others.

First, remember that you want an *exterior-grade* paint. That means it will stand up to inclement weather better than interior paint. Some climates require special paints that won't chalk (turn to dust) in heat and direct sun or that won't mildew in wetter climates.

The three most popular types of exterior paints are oil-based, latex, and varnish paints. What's the difference? The major difference is the primary ingredient, called the vehicle or carrier. Oil-based paints use linseed or a similar oil as its vehicle. Latex paints are water-based. Varnishes are solvent-based; they use mineral spirits, alcohol, or other solvents to carry the color.

As you can imagine, the primary ingredient or carrier is important to how the paint goes on and how it lasts. Oil-based paints and stains typically last longer than latex paints. However, products change, and today's latex is increasingly popular for new homes, once the domain of oil-based paints.

What's the difference between a paint and a stain? Stains have less pigment or color.

Ka-ching!

Most lines of paint include three quality grades: professional, standard, and premium. The best is usually the premium grade rather than the professional grade, which is typically made to be the least expensive, to keep costs down. Buy premium only, or make sure your painting contractor buys the best you can afford. It's more expensive, but will last much longer. Follow instructions on the paint can for cleaning up brushes, rollers, and other equipment.

What about equipment?

Most professional painters prefer sprayers for applying exterior paints and stains, and rollers to simultaneously smooth the paint (called back rolling). That's because it's faster to mask everything you don't want painted and to use a sprayer than it is to use brushes and rollers. It's time. And remember, time is money.

Most nonprofessional painters (such as owner-builders) prefer brush and rollers because equipment is less costly and less skill is needed. In either case, you can rent spray equipment or even professional painting equipment through most construction rental yards. So, here are some tips on applying exterior paints:

- Don't forget to apply a primer on siding, to seal the wood and prepare it for the paint.
- Only use exterior-grade paints or varnishes on the exterior of your house.
- Make sure the surface to be painted is clean (no sawdust) and dry, especially when applying oil-based paints.
- If using rollers, select ones made specifically for the type of surface you are painting.

- Invest in quality brushes; buy fewer and better ones, and take good care of them.

- Don't use natural bristle brushes for oil paints, as the bristles will curl.

- Small airless sprayers develop a paint fog rather than a spray, and are difficult to use for larger painting jobs.

Code Red

Professional sprayers use sufficient air pressure to infuse paint through the human skin! Be careful, and always use a facemask or respirator when spraying. Make sure you read and follow the safety guidelines recommended by the equipment manufacturer.

- Large compressed-air sprayers require that the paint be thinned out so that it will properly atomize or spray.

- Professionals often use airless diaphragm sprayers for high-volume painting with low overspray. Consider renting one, but make sure you get full instructions on how to use it properly.

- Power rollers are handy for painting large, flat surfaces fast.

- Apply paint evenly, and back-roll to remove any runs immediately.

- Rinse latex paint brushes with water if they will not be used again for more than two hours.

- Many paints are flammable; don't smoke or have an open flame near where you are painting.

Exterior Resources

Besides subcontractors and material suppliers, there are others who can help you finish the exterior of your home. They include masonry and siding specialists, roofing specialists, and painting equipment rental yards. Your local paint dealer may be helpful in selecting the best paint for the job, whether you do it yourself or hire it done.

In addition, *Do-It-Yourself Housebuilding* by George Nash (see Appendix B) is a recommended reference book aimed primarily at folks who want to do it all themselves.

The shell is finished. In the next chapter you'll start adding electrical, plumbing, and other vital services to your house.

The Least You Need to Know

- Select the best materials you can afford, and don't be afraid of doing the job yourself.

- Follow the building plans for installing a vapor barrier, sheathing, and siding.

- Even a small amount of masonry (stone, brick, blocks, stucco) can dress up the exterior of your home.

- Cornices or eaves are typically closed on conventional homes and open on nonconventional designs.

- Work safely, especially when painting the exterior of your home.

Chapter 22

Power to the People: Services

In This Chapter

◆ Reading the master service plan

◆ Materials you will need

◆ Installing electrical systems

◆ Installing plumbing and HVAC systems

◆ Installing other vital services

◆ Resources for services

What would life be like without electrical service, indoor plumbing, heating and air conditioning, and other household services? About where your house is right now: uninhabitable.

In this chapter I'll remedy the problem by showing you how electrical, plumbing, heating, air conditioning, cable, security, computer, and other services are installed in your home. Whether you're doing some or all of it yourself, or just want to make sure the contractors don't pull something over on you, this chapter offers the tools you need.

Checking the Master Service Plan

Your house plans will include either a master plan with utilities indicated or a specific set of service plans for electrical, plumbing, and so on.

That's your guide. For example, the electrical plan will show the location of the service entrance, meter, main circuit box, and circuits. Each circuit will include the location of switches, receptacles, and fixtures.

What the service plan *doesn't* tell you is how to get from here to there. That is, where should you run wires to furnish electricity to a light fixture through a switch?

Should you attempt to install your own electrical, plumbing, and other household systems? Good question! First, is it okay with the local building department? You probably asked this question when you applied for your building permit (see Chapter 11). Some local codes allow owner-builders to do their own electrical work (subject to inspection), but not the plumbing. Or vice versa. Check it out.

Typical electrical plan.

(© garlinghouse.com and homeplanfinder.com)

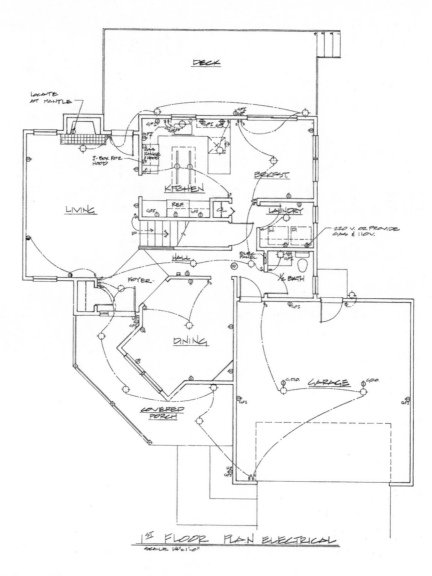

If it's okay with them, is it something you are qualified to do? That's a personal question. Before answering it, read this chapter through to see if you'll be comfortable with the job. Maybe not. Or maybe you will be, once you take a class or hire an advisor. You can also hire yourself out to the subcontractor as a helper, running wires or installing pipe as directed. It can both save you some money and help you learn a new skill.

There are two phases to installing most utility services: rough-in and finish. The rough-in is bringing the utility to the structure. Finishing means connecting it up. For example, plumbing is roughed in by getting the water and septic lines to the house and up into the walls per the plan. Finishing means bringing the pipes out of the closed walls and attaching them to sinks and other fixtures.

Although rough-in typically occurs before walls are closed and finishing after, both phases will be covered in this chapter for clarity. The demarcation between the two phases usually is closing up the walls. Before getting into the materials you need and the installation processes, let me offer some tips from the pros:

- ◆ You wouldn't attach a hose with the water on, so don't even try to wire your house with the electricity turned on.
- ◆ Make sure you use the size and type of materials indicated in the approved plans.
- ◆ White plastic (PVC) pipe is typically used for water and black (ABS) pipe for waste.
- ◆ Copper is better (and more expensive) than plastic water pipe.

◆ Read the fine print in your service plan. It will probably show the wire gauge that is approved for installation.

◆ Even if you plan to run the wires from the receptacles to the service panel yourself, hire an electrician to install and prep the service panel.

◆ If you make any changes to the service plan, make sure the changes get marked on all relevant copies.

Selecting Utility Materials

Your house's service plans and materials list will outline the specific materials you will need. But let's discuss some of them here.

Electrical Materials

The primary material used in electrical systems is wire. You knew that. In the old days, individual wires were run through walls, kept away from wood and other flammables with ceramic insulators. Today's wiring systems are wrapped in insulation and wrapped again in plastic. It's commonly called Romex (pronounced *RO-mex*; a brand name), but it's actually type NM cable. NM simply means nonmetallic, because the cover is plastic rather than metal. Two or more wires inside are insulated with thermoplastic (T wire) and the ground wire is bare. The package is sheathed in plastic. There are also TW, THW, NMC, UF, and armored cables for specific applications.

Wire size is indicated by its gauge. *The smaller the gauge number the larger the wire, and the more electrical current it can carry;* 14 ga. wire is *smaller* than 12 ga. An electrical cable labeled *Type NM 12-2 w/G* has a non-metallic (plastic) sheathing covering two 12 ga. wires plus a ground wire. Again, your electrical service plan will indicate what wire gauge is approved for each application.

You can't use smaller (higher gauge) than approved, but you can use larger (lower gauge) wire. Why should you? Safety. Larger wire can carry more current with less heat. Check local prices, but for a few dollars more, 10 ga. or even 8 ga. wire is a good investment.

Most of your house will use 110-volt electrical service (110v) for things like lighting and most appliances. Electric dryers and some shop equipment may require 220v service. This chapter talks about 110v.

Wires come in colors to tell you what they're for. Wiring systems need a "hot" wire and a neutral wire. Most also have a ground wire. The voltage difference between the hot and neutral wire (110 volts) is what drives the electrical current. Black, red, and blue wires are "hot," white wires are neutral, and green, green-yellow, and bare copper wires are ground. Those are probably all the colors you'll run into.

Who cares? If wire is wire, why not use whatever color you want? For consistency. Five years from now, you're installing a new light fixture and don't know which wire is hot and which is ground by looking at them—except by the color. Meantime, your building inspector will care! So don't try to make a fashion statement by mixing wire colors.

Code Red

If you're using aluminum wire for a circuit designed for copper wire (not recommended), use aluminum wire two sizes larger (that is, two gauge numbers smaller). Replace 12 ga. copper wire with 8 ga. aluminum wire. Aluminum isn't as conductive as copper.

Some building codes call for shielding the wire with metal conduit, flexible pipe through which the wires are run. If your house plans require some conduit, make sure you install the correct wiring in it. NM or Romex isn't installed in conduit.

Electrical fixtures include receptacles (plugs), lighting fixtures, and switches. Each requires that a plastic or metal box be installed in the wall (before drywall) and wires run to it. Once the wall is closed, the receptacle, fixture, or switch is connected to the wires and attached to the box.

Each circuit is designed (and approved) to carry a specific amount of electricity and no more—hence circuit breakers that turn things off when too much juice flows. What you need to know is the amperage (amp) rating for each circuit. That number dictates the wiring gauge and circuit breaker used in the circuit. For example, lighting circuits are typically limited to 15 amps (15a). Those that serve small appliances can go up to 20a.

Plumbing Materials

I doubt that any owner-builder has ever drowned by installing plumbing. So it's safer than installing electrical systems. However, many building codes will let owners install electrical systems, but not plumbing. Go figure. In any case, the primary material in plumbing is—surprise!—pipe.

Plumbing delivers fresh water and removes used water from your home. The fresh water is delivered through the supply line. The used water leaves via the DWV (drain-waste-vent) system. Don't get the two confused as you plumb your house!

There's another important system, too. Actually, it's a sub-system of the supply line. It's a hot water system that heats and delivers supply line water.

Supply line pipe is made of steel, copper, or plastic. DWV systems use plastic or iron pipe. Plastic pipe is actually PVC, CPVC, or PB (acronyms for unpronounceable terms). PVC and CPVC are rigid. CPVC is better for hot water pipes.

Local code and your house designer will indicate on the plumbing plans what material you should use for your pipes. If a preference is available, copper supply pipe typically is preferred over plastic, and plastic DWV pipe is preferred over iron. Plans will also indicate the required size and location of pipes, making the job easier.

What becomes complex in plumbing is figuring out and installing traps and vents. Simply, a trap keeps yucky gasses in the DWV system pipes from coming into the house. The water in the bottom of a toilet is a built-in trap. Vents allow the yucky gasses to vent out an open pipe (stack) through the roof.

Water systems get pretty complex as you try to pull everything together: supply line, hot water, cold water, softened water, DWV, traps, vents, revents, cleanouts. Whew! Lots of pipes running in all directions! Thankfully, you have a plan—a plumbing plan.

Other Service Materials

Your home may have other services including heat, air conditioning, telephone, cable, satellite, music, intercom, doorbell, smoke alarm, burglar alarm, computer network, vacuum system, and more. (Wouldn't it be great to install a system for delivering beverages from the refrigerator to your easy chair?! If it's not on your building plan, you can still add it in before the walls are closed up.)

Most of these systems have three things in common: input, control, and output. The telephone wires come into the house, are attached at plugs, and the phones are plugged in. A burglar alarm system requires electricity to run it, a sensor triggers the alarm, and it tells someone audibly or electronically.

So installing other systems means selecting the components, figuring out where they will go and what they need to get there, and then actually doing the work. A doorbell, for example, needs low-voltage electrical wire, a transformer, a bell, and a button. Install the button at the front door and/or other locations; install the bell where you can hear it throughout the house.

The largest optional system is heating and/or air conditioning, typically called HVAC (heating, ventilating, air conditioning). It, too, has inputs (electricity, air, fuel), controls, and output (hot or cold air). The system may be as simple as radiant-heat baseboard heaters or as complex as a full heat-pump system.

Though do-it-yourselfers typically don't install furnace or air conditioning systems, they can help to install the ducts that channel air to and from the systems. Follow your building plans and detail sheets for selecting and installing these systems.

Installing Electrical Service, Lighting, Appliances

Roughing-in an electrical system means running wires from the service panel to the electrical boxes. Finishing requires connecting the wires and fixtures.

Electricity comes into the house from the service line, through the meter, and into the main service panel. The panel includes a main disconnect switch to stop all electrical service to the house (a handy gadget in an emergency) and individual breakers that control electricity to specific circuits. The box will also have a single wire connected to a nearby water pipe or metal stake that serves as the system's "ground."

The service panel is wired so that all the neutral wires are tied together, all of the ground wires meet at a *bus bar*, and all the hot wires are attached to their respective circuit breakers. Circuit breakers break or stop the circuit (flow of electricity) if the rated amperage is exceeded. Good idea!

The service panel distributes both 110v and 220v electricity. The 220v system uses two hot wires instead of one to double the voltage. Some electrical plans call for a second service panel somewhere in the home, typically near a large circuit such as HVAC, utility, or garage.

Building Your Vocab

Bus bar is a common point where all wires with the same function meet, such as a ground bus bar or a neutral bus bar. It's also where single busses meet on a Friday night.

Most electricians install electrical boxes before installing wire. Then they know how long each of the needs to be.

Electrical cable is relatively cheap. If you're not sure how much you need, cut each cable 5 fe 10 feet longer than needed so you won't have to splice or replace the cable. Sure, having l wasted cable lying around looks amateur, but so does installing an unnecessary electrica the run.

Installing wire cable between an electrical box and the service panel is called pull and fixture boxes, and then drill holes in studs along the most direct path to th lights and switches typically run higher in the wall and in the ceiling. Cabl wall, usually about 12 inches above the floor. Once you cut the length a what it goes to. You then "make up" the service panel by attaching th marking them.

Electrical service panel.

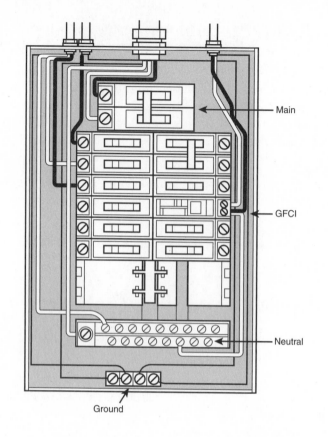

Once the cables are in place, remove slack and (carefully) staple the cable to wall framing so that it doesn't move around. Your house's electrical plans and building department can tell you how frequently staples are needed (typically, every 4½ feet, and within 12 inches of the electrical box).

You install electrical boxes and run cable from the service panel(s) to the boxes. Good work! You can finish the job once the walls are insulated and drywall is installed. Chances are, the building inspector will want to check it out before you proceed.

> ### Ka-ching!
>
> Save yourself lots of time and frustration by marking each circuit. Lay a piece of masking tape next to the circuit breakers in the service panel and put a number by each breaker. Then wrap a piece of masking tape at the other end of each cable with the appropriate number. Once the walls are closed up, it won't be as easy to trace a cable back to find out which circuit it's on.

Finally, install the receptacles, fixtures, and switches according to package instructions. Most are straight-forward. The only tricky wiring is for multi-wire circuits such as multiple switches that control the same fixture. Here's a hint: In most cases, wires are connected by colors. Black-to-black, green-to-green, ground-to-ground. If in doubt, get help. Be aware that an electrical inspector will more carefully study an owner-builder's work than a licensed electrician's. If you have questions, you can call your inspector in advance. Consider asking for (and paying for) an interim inspection to make sure you're on track.

ing Plumbing and Related Services

sn't so bad! Now, if you can do your own plumbing, here's an overview. If nothing else, it offers you keeping your plumbing contractor relatively honest.

Just as with electricity, the house's water system comes in from the street (or from your well or spring). There's a water meter so that the water company knows how much you've used, and charges you accordingly. A shutoff valve is installed in case you need to turn off the water system, such as during construction. Then a service pipe feeds water to your house. (You may have a stop-and-drain valve located here.)

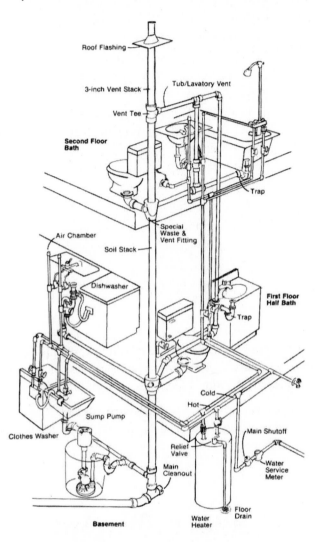

Cross-section of a typical home's plumbing system.

Pipes are made of galvanized (zinc-coated) steel, copper, or plastic (PVC, CPVC, PB). Galvanized pipe is threaded to screw together. Copper pipe is soldered together. Plastic pipe is glued together with a special solvent. Smaller plastic lines are threaded to screw together. Check your local building code to see what's allowed, and check with your material supplier for relative costs.

What size pipe? Check the plans. Pretty common is 1" supply for hot and cold mains; ¾" distribution lines; and ½" lines to fixtures. Drains are usually 4"; some kitchens have a 2" drain. Bathroom sink and tub drains are typically 1½". Vents are 2". Your codes may vary. Your house's plumbing plan offers the specifics on where everything is located. It will be up to you (or your plumber) to get it all connected up and operating. Here are some tips:

- ◆ Make sure that all pipes are installed either horizontally or vertically. Pipes that run at angles aren't efficient.
- ◆ Follow the plans to include traps and vents in DWV lines.
- ◆ As possible, include cleanout plugs in DWV lines and traps.

- The easiest way to run pipes from room to room is through the floor, and then up through walls. Add pipe supports as needed.

- Include an air chamber above lines into faucets to prevent air pockets from rattling pipes, called water hammer.

- Wrap Teflon tape around threads to seal joints.

- Plan for your water heater to be centrally located to bathrooms and kitchen. If necessary, install a smaller second water heater upstairs.

- Pick up free brochures and instruction sheets at your building material supplier on how to install plumbing and fixtures.

- Remember that plumbing fixtures (toilets, tubs) located in a basement and below the incoming sewer or septic line will require special fixtures.

- The trickiest plumbing job is installing a one-piece shower/tub in the bathroom. Make sure everything is measured out and double-checked before installation.

- Platform tubs and most sinks require some type of framing to support them.

Ka-ching!

Many do-it-yourselfers prefer copper or plastic pipe, because galvanized pipe requires expensive tools for cutting and threading. Plastic pipe is the easiest to install. Copper is typically the best choice—and the most expensive.

Installing HVAC Systems

There's only one primary way to distribute electricity or water within your home, but there are lots of options for keeping your home comfortable. You can use fossil fuel, solar, electric, hot water, or a heat pump to heat it. You can use air flow, ventilation, berms, evaporative coolers, and air conditioning to cool it. And you can distribute the treated air naturally, with ceiling fans, or with forced-air systems.

The smart owner-builder who has a pro designing the system and advising on its installation can tackle most of those options. You can purchase furnaces and air conditioners through building material suppliers. Handy people can buy and install rigid and flexible ducting. You can hook up controllers to existing electrical systems. The best advice is either to have a pro install your HVAC system or hire one to at least select it and advise you on its installation.

On the Level

If you're interested in including solar power or other alternative energy sources in your new home, read my book *The Complete Idiot's Guide to Solar Power for Your Home* (see Appendix B). It covers practical tips on selecting, sizing, financing, installing, and using solar, wind, and other power sources for better living. It's available through local bookstores or at www.MulliganBooks.com.

Installing Other Systems

As I mentioned earlier, there are many other features you can add to your home. They include telephone systems, built-in vacuum system, cable, satellite, music, intercom, doorbell, smoke alarm, burglar alarm, computer network, and others.

Knowing what you now know about running wires and pipes through walls, you can see how computer network cable, a security system, or a built-in vacuum system is installed. Centralize the primary equipment and run wires, conduit, or pipe through walls and floors to sensor and controller locations.

Most important, follow your approved building plans.

Service Resources

Depending on how much of your home's services you plan to install yourself, you may need more help. The National Electrical Code (NEC) is developed and published by the National Fire Protection Association (NFPA) and is available through larger bookstores or online at www.nfpa.org.

The National Association of Plumbing-Heating-Cooling Contractors (NAPHCC) publishes the National Standard Plumbing Code (NSPC) available through bookstores and online at www.naphcc.org.

In addition, your local building department can recommend specific books and pamphlets to help you through the specifics. If all else fails, hire help. It's cheaper to have someone do a job to code than to redo a job to bring it up to code.

Next, let's finish up the inside of your home!

The Least You Need to Know

- Your house's master service plan should include the specific details of materials needed to install electrical, plumbing, HVAC, and other services.
- The rough-in is the first step to installing utilities, and it precedes closing the walls.
- Finishing electrical, plumbing, and other systems starts once walls are closed and the building inspector has approved prior work.
- Many other useful systems (telephone systems, built-in vacuum systems, cable, satellite, music, intercom, doorbell, smoke alarm, burglar alarm, computer network) can be roughed-in now and enjoyed later.

Chapter 23

Somebody Get the Door: Interior

In This Chapter

- Selecting interior materials
- Installing walls and ceilings
- Installing doors and windows
- Installing floors and cabinetry
- Resources for interiors

There have been many changes in your lot over the past few weeks! It now has a foundation, platform floor, wall framing, exterior sheathing, roofing, roughed-in electrical and plumbing, and more. You've been busy!

And you're going to get even busier in this chapter. You'll be insulating and closing up walls and ceilings, installing doors and windows, and adding floors and cabinetry. You'll learn how to determine the best sequence for finishing your home, called critical path. So grab another double mocha and let's head out to the job site. (Remember to check Appendix A for the definition of any unfamiliar building words.)

Selecting Materials for the Inside

What materials will you need to finish the interior of your house? Your materials list will give the specifics. For now, let's first review the process. Then we'll consider the primary interior materials used in constructing your house. Here's what you'll be doing:

- Installing insulation material in walls, ceilings, and floors.
- Making sure that the walls and ceilings are ready to be closed up.
- Installing drywall, plaster, and/or paneling.
- Installing windows and doors in frames, and then finishing with trim.
- Preparing the subfloor and installing flooring or carpeting.
- Installing kitchen, bath, and other cabinets and countertops.
- Finishing up electrical, plumbing, and other services (see Chapter 22).
- Installing interior masonry, stone, and/or marble.

The primary materials you'll need are the insulation, wall covering, flooring, doors, windows, and cabinets. Selecting insulation was covered in Chapter 6.

Wall-covering materials include drywall, plaster, and paneling. Drywall, also known as gypsum wallboard panels or Sheetrock, is by far today's most popular wall finishing material. It's inexpensive, relatively easy to install, fire resistant, accepts paint well, and you can use it for both walls and ceilings.

Drywall panels are sold in sheets of 4' × 8' to 4' × 12'. Professionals prefer the larger panels because they require less finishing. Owner-builders typically prefer the 4' × 8' because they're easier to handle. They weigh about 60 pounds per sheet.

Standard drywall thicknesses are ¼", ⅜", ½", and ⅝". Your local building code may require a specific thickness and fire-code rating. The most commonly used thicknesses are ½" for walls and ⅝" for ceilings. Ceiling panels are thicker because they can sag. You can cover curves in walls and ceilings with two overlapping layers of ¼" drywall.

Drywall panels have tapered edges. Once two panels are butted on a wall, you cover the indentation formed by the tapered edges with a special drywall tape and fill with a thin layer of plaster to smooth the surface.

Ka-ching!

Plan out your drywall in advance. One edge of drywall panels isn't tapered, making taping and smoothing it more difficult. Plan installation so that the untapered edge is at right angles to another panel, such as at the junction of two walls or a wall and a ceiling.

Drywall is attached to walls and joists using special drywall nails or screws. Screws are preferred because they hold better, but special drywall nails with ring shanks do a good job. Adhesives can help keep panels in place. Most owner-builders and pros prefer screws because they're faster to install. You'll learn how to install drywall panels later in this chapter.

Some folks still prefer plaster walls. They require a backing such as lath or perforated panels to which wet plaster is spread and smoothed. They also require special skills that most owner-builders don't have.

Paneling is relatively easy to select and install. Depending on local code, you can install decorative paneling of wood or other material directly over wall framing. Others will require a subsurface of drywall for fire resistance. Make sure the paneling you select meets local code requirements.

Flooring materials include carpet, solid wood, wood laminates, hard tile, and resilient tile. Owner-builders prefer flooring that is easier to install, such as wood laminates and resilient tile. Carpet comes in 12'-wide rolls. Wood floors are sold by the row to cover a specific square footage. Tiles come in 10" × 10" or 12" × 12" squares.

Carpeting typically is priced by the square yard (sq. yd.; 9 sq. ft.) and is in the middle of the cost curve. Less expensive are wood laminates and resilient tile. Above it, cost-wise, is hard tile followed by solid wood flooring. However, if you shop around you can get more for each dollar you spend on flooring.

Windows come in a wider variety of sizes and styles, as shown in the following illustration. Double-hung windows have a stationary top panel and a sliding bottom panel. Gliding windows have one stationary and one sliding side panel. Casement windows swing out from hinges at the side. Awning windows swing out from hinges at the top. Bay windows have three sides with sharp angles. Bow windows have more than three panes and a smoother angle. Choose window designs that best fit the style of home you are building.

Windows have become very energy efficient over the past few decades, with some modern windows nearly as efficient as the wall they replace. How energy efficient yours need to be depends on where you live. Ask your materials supplier and/or contractor to help you select cost-effective, energy-efficient windows and doors.

Windows are priced by the size, number of panes of glass, and whether there is an additional operating mechanism. A standard-size, single-pane, double-hung window is much less expensive than a custom double- or triple-pane window with crank-out. Cost-wise, the price of windows goes up from wood to vinyl to aluminum depending on the quality of the materials and the workmanship.

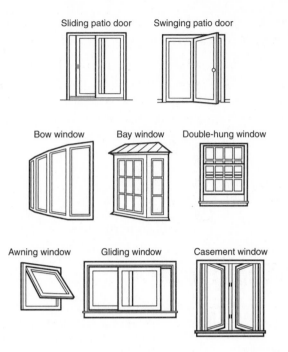

Common window styles.

Cabinets are designed with your home and are ordered as stock or custom cabinets, or both. Standard kitchen base cabinets are 36" high (including base and top) and 24" deep. Width depends on the cabinet's function and whether it will have doors or drawers. Common widths of lower or base cabinets are 18", 24", 30", 36", 42", and 48", all increments of 6". Upper or wall cabinets are typically 30" high and 18" over the top of base cabinets. Modular upper cabinet widths are available in 9" increments.

Cabinets are priced by the size, wood, and number of drawers or doors. If possible, buy cabinets of standard heights, widths, and depths that can be mass-produced rather than custom cabinetry that can cost twice as much or more. In addition, solid wood cabinets are more expensive than those that use cheaper woods covered with quality woods.

CAUTION

Code Red

Installing nonstandard height kitchen cabinets (for taller or shorter people) can adversely impact home resale. However, if you plan to live in the home for a long time, size them to fit you. The base or counter top can be modified later to make it standard.

Insulating Your Home

Insulating walls and ceilings in your home is relatively easy. (Remember that you should rough-in all electrical, plumbing, and other services, as described in Chapter 22, before installing insulation and closing up walls and ceilings.) Most insulating materials today are designed to both do the job and make it easy to install. Insulation rolls are made to fit between wall studs. Alternatively, loose insulation is blown into wall cavities.

The most popular type of insulation in conventional homes is rolls of woven glass or mineral fibers glued to a paper backing. Here's how to install:

1. Unroll and cut to the height of the wall cavity.

2. Turn the backing paper toward the inside of the house and press the batt into the cavity.

3. Staple the edge of the backing paper to the wall studs to hold it in place.

Special blown-in insulation adheres to the wall cavity, but requires special equipment for installation. Rigid insulation is installed either on the interior or exterior of exterior walls. The process is the same as installing exterior sheathing (see Chapter 21).

Ka-ching!

As you are installing insulation, look for openings in the wall construction where air can leak and seal them. A few extra minutes doing this can save you hundreds of dollars in energy costs over the years to come.

Some building codes and approved building plans will require installing a moisture or vapor barrier. The barrier is typically 2- to 4-mil-thickness plastic film stapled across interior studs on exterior walls. You can install vapor barriers on interior walls as well, especially bathroom and kitchen walls where moisture develops.

Alternatively, you can install building felt or other materials as a moisture barrier. Floors are typically insulated with batts of fiberglass insulation. Insulate ceilings with fiberglass batts or with loose-fill insulation following manufacturer's instructions.

Installing Drywall and Plaster

Drywall is really handy stuff. It's dry plaster wrapped in paper. It makes covering a wall much easier than wet-plastering. But it's still a job. Here are some tips for installing drywall in your house:

◆ Install drywall on the ceiling before installing it on walls.

◆ Start installing drywall panels in the middle of the ceiling and work toward the edges.

◆ Use a T-shaped cradle and/or a helper to hold drywall panels in place while being fastened.

◆ Power screwdrivers make fastening drywall panels easier and faster than manual nailing.

◆ Fasten the center of the panel first, and then work toward the edges.

◆ Use special drywall screws or nails that imbed into the wood at least one inch.

◆ Stagger drywall panel seams so that the joints are as short as possible.

◆ Whether nailing or screwing, slightly depress the fastener's head below the panel surface (called a dimple); it will be leveled with plaster later.

◆ Nail drywall panels horizontally on walls, starting in a corner.

◆ Install the first wall panel's edge against the ceiling.

◆ Cut and install the second wall panel below the first, trimming the lowest edge if needed.

Drywall panels are easy to cut. Measure and mark, and then use a straight-edge and a drywall knife or box cutter to cut through the panel's paper. Then apply pressure on the cut to break the panel. Finally, use the knife to cut through the paper on the other side of the panel. Once all the panels are up and fastened, they are taped and cornered. Taping drywall means filling indentations, including tapered edges and nail dimples, to make the surface smooth. Here's how it's done:

1. Spread a thin coat of joint compound (a fine wet plaster mix) over joints with a joint trowel.

2. Spread joint tape on the compound along the joint, using the joint trowel, and allow it to dry.

3. Apply a second thin coat of compound wider and allow it to dry.

4. Apply another thin coat, even wider, and allow it to dry for 24 hours.

5. Cover all fastener dimples with a thin coat of joint compound.

6. Smooth the finished joint with a damp sponge and let it dry before painting.

7. If sanding is needed, use a smoothing screen sander available from your hardware supplier rather than sandpaper.

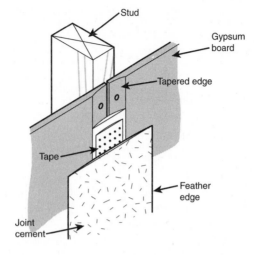

Taping drywall.

Drywall wall corners are delicate. Bumping into a corner can damage it. The solution is to install metal corners. These are special metal strips that fasten to the two drywall panels to protect the corner. Corners are cut to length and then fastened to the panels. Finally, joint compound is applied to make the transition smooth.

You may finish off your drywall with a texture or with paneling. Texturing takes some experience to do right. It's done by applying a thicker plaster material, and smoothing out the highest points with a large trowel. Texturing is applied to ceilings using a special sprayer, requiring experience to do it right. You can install paneling using adhesives. I'll cover this topic in greater detail in Chapter 25.

As I mentioned earlier, installing a plaster wall requires a backing. It could be lath (wood or metal strips) or perforated panels nailed to studs. In fact, some backing panels for plaster walls are made from dry plaster similar to drywall. Perforations (and space between laths) give the wet plaster something to hold on to. The final coat is smoothed with a wet trowel.

Code Red _____

If you'll be doing any sanding, always use a respirator or wear a safety mask to avoid inhaling plaster.

Stucco interiors are installed like plaster, except over a wire mesh attached to wall and ceiling framing. It's not the typical job for do-it-yourselfers, but can be worth the extra cost of hiring an experienced stucco tradesperson.

Installing Doors, Windows, and Trim

The primary exterior door is typically 36" wide. Most other exterior and most interior doors are 30" wide. Double-entry and sliding glass doors are 60" or 72" wide. Standard door height is 6'8".

Doors, windows, and other barriers are installed either after the exterior is sheathed (see Chapter 21) or after the interior walls are closed, as described here. Fortunately, modern doors and windows are pre-hung. That is, they are purchased as a package that includes casing and even trim, making installation easier. In most cases, all that's needed is to place the unit in the wall opening, make sure it is level and square, and fasten it in place.

Also, windows and doors are standardized to ensure that they are energy efficient, well built, and will have a long and functional life. So, let's start installing them. Here are the typical steps for installing a standard pre-hung door:

1. Verify that the rough opening for the door is correct (instructions with the door offer specific dimensions).

2. Make sure that the opening is square and level.

Tilting the pre-hung door into place.

3. Unpackage the door assembly and remove the skid (bottom) board used for support in transit.

4. Based on manufacturer's instructions, prepare the door's threshold now or once the door is installed. Also remove any trim required by the instructions.

5. Center the assembly in the opening and tilt it up into position.

6. Plumb the door frame on the hinge side, using screws to temporarily fasten the frame to the stud.

7. Plumb the door frame on the latch side, using screws to temporarily fasten the frame to the stud.

8. Insert wedges or shims (typically cedar or kiln-dried wood) behind the hinges and other points around the frame for support.

9. Make sure the door frame is plumb and square, and then finish attaching the frame to the studs by the manufacturer's instructions.

10. Remove any shipping clips that hold the door closed.

11. Install and/or adjust door locks and hardware.

12. Install insulation as needed between the opening and the door assembly.

13. Install trim.

14. Apply caulking around the door.

15. If not previously required, install the threshold.

Exterior and interior doors are hung in approximately the same manner. The major difference is that exterior doors are typically wider and heavier. Local codes may require a specific type or rating of door. Code is a minimum, not a maximum, so follow it. Patio doors are larger and require more help, but are installed in a similar manner as standard doors.

Here are instructions for installing standard pre-hung windows:

1. Verify that the rough opening for the window is correct (instructions with the window offer specific dimensions).

2. Make sure that the opening is square and level.

3. Unpackage the window assembly and remove any components used for support in transit.

4. Remove any trim required by the instructions.

5. Center the assembly in the opening and slide it into position.

6. Plumb the window frame on all sides and use screws to temporarily fasten the frame to the stud.

7. Insert wood wedges or shims around the frame for support.

8. Make sure the window frame is plumb and square, and then finish attaching the frame to the studs by the manufacturer's instructions.

9. Remove any shipping clips that hold the window closed.

10. Install and/or adjust window hardware.

11. Install insulation as needed between the opening and the window assembly.

12. Install trim.

13. Apply caulking around the window.

Skylights are simply windows in the roof. Installation is similar to that for windows, with a few exceptions. Most important, make sure that the seal between the frame and the skylight is adequate. Also, following the manufacturer's instructions, install metal flashing around the skylight.

Door and window trim may or may not be included in the pre-hung assembly. If not, or if you'd prefer trim that better matches the interior of your home, it's time to install trim. Depending on the type of doors and windows you install, you may also need to make your own casings or edge trimming. Door and window trim is typically cut at an angle. This is where a miter box is very useful. You can buy or rent a trim box that includes the miter box and special saw for fine cutting.

Installing Finish Flooring

Chapter 19 covered the installation of your house's platform or subfloor. But the floors were ugly. Now you'll finish them with wood, tile, carpet, or something that looks like one of these materials.

First, consider *when* you should install the finish flooring. Many contractors install finish flooring after everything is painted and just before the plumber installs (sets) the toilet. It's one of the last things on the list. Owner-builders may opt to do the same or to install hard flooring now and any carpeting later. Just make sure it is installed when it makes the most sense for your project.

There are many laminate and tile floor products on the market designed specifically for installation by do-it-yourselfers. They include laminate wood floors that interlock and glue to the subfloor. There are even ones that don't require glue!

> **CAUTION** **Code Red** _____
>
> You can probably install your own carpeting if needed. However, it's usually wiser to hire an experienced carpet layer and try saving money elsewhere. Why? Because the trick to laying carpet is installing it with a sturdy and invisible seam, and that can be difficult for a novice to do.

Installation of most of these floors is similar. The subfloor is cleaned of any rough spots including plaster and protruding nails. In some cases, a vapor barrier or a soft underlayment is installed. Then the flooring is installed with glue or fasteners.

Tile floors typically require a mastic or thick glue over which the tiles are placed. Plastic spacers keep the tile at the required distance from each other for grout. Then grout (a sandy mortar) is spread between the tiles and smoothed out. Some grouts are then sealed.

Installing Cabinetry and Countertops

Here comes the cabinet truck! It pulls up and drops off all the cabinets you previously ordered, ready for installation. If you've hired a carpenter to set the cabinets for you, there's nothing more for you to do except supervise. But if you're going to set them yourself, here's how it's done:

1. Measure and mark the floors and walls so that you know exactly where all cabinet components will go.

2. Set all the cabinets in their respective places to make sure you know you've got everything and that everything fits.

3. Remove cabinets from the area as needed and begin installing or building the bases.

4. Make sure the bases are level and square, and then install them permanently.

5. Install the base cabinets on top of the bases, making adjustments as needed for pipes and wiring.

6. Plumb and install the wall cabinets, screwing them to wall studs and to the soffit.

7. Install the counter top and splashguard.

8. Install sinks, faucets, and other fixtures according to the manufacturers' instructions.

Bathroom cabinets and vanities are installed in the same way.

On the Level

Hard interior surfaces, like masonry, stone, and marble, typically require skills beyond that of most do-it-yourselfers. If your budget allows, hire an experienced contractor or sub-contractor to do the work for you. Or check out the many good books available on stone-work, laying marble, fancy brick-work, tiling, and other skills.

Interior Resources

The best resource for planning, contracting, and installing interiors is your building materials supplier. A few hours wandering around a large store can give you dozens of ideas on what you want and how to install it. And don't by shy. Ask questions. You may find a clerk or two who has worked in the trade and can give you real-world answers to your questions.

You can find other useful books with specific information on various building skills. One that I recommend *How to Design & Build Your Own Home* by Lupe DiDonno and Phyllis Sperling (see Appendix B). It emphasizes what you need to know when designing your house.

Your home is coming along quite nicely! In the next chapter I'll give you some tips on decorating options for your home before you move in.

The Least You Need to Know

◆ Drywall is the most popular wall and ceiling material because it is relatively inexpensive and easy to install.

◆ Pre-hung doors and windows make installation easy.

◆ There is a wide variety of flooring materials that you can install, and many others that require professional experience.

◆ Standard cabinets are easy to install with basic tools and rules.

Chapter 24

Decorating Finishing Touches

In This Chapter

- Decorating your new home: Do it yourself or hire a pro?
- Trimming and painting the interior
- Installing lighting and appliances
- Selecting furniture and finishing touches
- Resources for decorating

Now that your home has been built and finished inside and out, it's time for the fun part! Your new home is a palate just waiting for you to add those individual touches that will make it uniquely yours.

This chapter offers dozens of ideas on how you can enhance the personality of your new home with paint, trim, lighting, appliances, furniture, and other elements. So open up your Home Book and let's start decorating your new home!

Making Your House a Home

Like every aspect of building a new home, you have options for decorating it. You can hire a professional *interior designer* to advise you or to surprise you. You can try your hand at decorating your own home. Or you can simply move everything in from your last residence and take your time discovering the personality of your new home.

If you decide to get some professional decorating help, you still have some choices. You can hire an interior designing firm that will use on-staff specialists to tackle specific challenges. You can hire an independent *interior decorator* who will do all the design for you. Or you can work with a decorating or furniture store that will sell you what you need.

Your choice among decorators depends somewhat on your budget as well as on your decorating challenges. Decorating a log home, for example, may be easier than tackling a custom-designed home. Log homes tend to have colors and furnishings that fit into a specific category: Earth tones and mostly wood furnishings give them a rustic appeal. Custom homes, on the other hand, are blank palates to which you can add a colorful personality.

Interior decorators are paid either by the hour or by a percentage of what's purchased. Independent decorators, for example, may get 25 to 40 percent of your decorating budget in exchange for their services. However, they should also pass along any discounts they earn from retailers. The decorator may also get her or his fee directly from the retailer. Depending on the size of your budget, you may get a small discount as well.

If you're considering working with an interior designer or decorator, here are some questions you should ask:

◆ What do you specialize in?

◆ What professional certifications do you have?

◆ Can you tell me about some projects you've completed?

◆ Is my budget realistic? Why or why not?

◆ What are your fees and who pays them?

◆ What discounts are available to me?

◆ How much time should be planned for this project?

As you decorate your home, remember that it's just stuff. Don't be overly concerned about selecting just the right tone or the perfect accent. Your home doesn't have to resemble a showplace or something out of a magazine. Have fun! Surround yourself with pieces that you like and that work with your lifestyle. You can always change them later as your tastes change.

For additional ideas on hiring professionals, refer to Chapter 15.

If you are decorating your own home, develop your knowledge of residential design through magazines and books. Consider taking a class or two. Become more aware of colors and other design elements as you visit homes and professional offices. Also, find local paint and furniture stores that have knowledgeable staff to advise you.

Beautifying with Trim

Trim is a necessity. It covers the rough edges of construction. Trim is applied around doors and windows to cover the frame. It also covers the edge between the walls and floor.

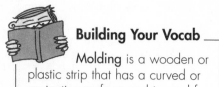

Trim is also a luxury. Special trim such as wainscoting decorates walls. Ornate ceiling molding adds beauty to a room. Some pre-hung doors and windows include standard trim *molding*.

In addition, trim is relatively inexpensive, costing under one dollar a running foot for many popular types in common woods. It gets more expensive as you select hardwoods and exotic woods. The cost also goes up for larger pieces or ones that are more ornate. You can save money if you hand-select the best trim and finish it yourself.

Molding is relatively easy to install. Only basic tools are needed:

◆ Finish hammer and nail set

◆ Miter box

◆ Miter, back, or coping saw

◆ Finish nails

Exact installation depends on the design of the molding. Simple door and window molding simply butts a vertical piece against a horizontal one. No special angles to cut. Corner molding typically requires an angle cut; that's where the miter box comes in.

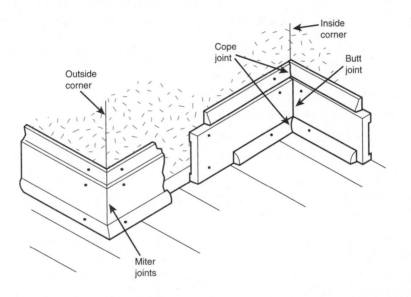

Typical baseboard trim.

A miter (pronounced *MY-ter*) box is a tool that enables you to make perfect angle cuts. Wood and plastic miter boxes are adequate for simple angle cuts. If you plan to do more complex angles, opt for a steel miter box or even a power miter box. Here are some tips on cutting and installing trim molding:

◆ Practice cutting and fitting trim on scrap wood.

◆ Attach the miter box to a rigid surface.

◆ Make sure your saw is sharp and the sides are smooth so that it tracks easily in the miter box notches.

◆ Start with the easiest trim and work up to the pieces that are more difficult.

◆ Apply the first coat of paint to cut trim, and the second or touch-up coat once it is installed.

◆ Splice longer trim runs by cutting both pieces at an angle.

Painting the Interior

Interior painting is a relatively easy job that many owner-builders tackle themselves. It can be done prior to occupancy, after moving in, or both.

Both? Yes, you can paint the interior of your home a neutral color for now and plan to repaint specific rooms later as you determine what colors you'd like there.

On the Level _____

When should you paint the interior of your home? That depends on who is doing the painting and the other jobs to finish your house. If hiring contractors, interior painting and trim should occur prior to floor covering so that they don't walk on or spill on the new floor. In this case, the only workers who may walk on your new floor are the plumber setting the toilet and electrician hanging a lighting fixture. If you're doing it all yourself, it depends on what's most efficient for you.

Paint has been enhanced by technology. Today's paints are of better quality, longer life, and greater value than those of just a decade ago. They are also easier to apply. I described exterior paints in Chapter 21. Interior paints are similar, but they have to stand up to kids rather than to the elements. So, remember to purchase _interior grade paint_ rather than exterior.

Like exterior paints, interior paints come in either oil-based, latex, or varnish varieties. And like exterior paints, the major difference is the primary ingredient, or carrier. Oil-based paints use linseed or a similar oil as its carrier. Latex paints are water-based. Varnishes are solvent-based; they use mineral spirits, alcohol, or other solvents to carry the color or pigment.

Which type of paint you use depends on what characteristics you want. Latex paints are easier to apply and to clean up. Oil-based paints offer resistance to wear. Varnishes stand up well to wear while retaining the beauty of underlying wood.

Latex paints are a popular choice for the majority of interior walls and ceilings. Because of this popularity, many varieties are available for special applications:

♦ **Flat** latex paints allow the surface to breathe without sealing it. This is a good feature for walls and ceilings in the main living areas.

♦ **Gloss** latex paints seal better against moisture and wear. They are popular for bathroom and kitchen walls and ceilings. They are also used on trim, especially doors and windows subject to handling and wear.

♦ **Semi-gloss** latex paints offer some of the benefits of both flat and gloss paints. They allow surfaces to breathe while making cleaning them easier. Semi-gloss paint is a good choice for bedrooms.

As you can imagine, the line between these types of latex paints is pretty loose. In fact, one brand's semi-gloss is another company's quality flat. Visit local paint stores for more specifics on which type paint suits your home's needs.

Paint is applied using brushes, rollers, and sprayers. Typically, professional painters prefer sprayers because they are more efficient—if you know what you're doing. So-called airless sprayers sold to nonprofessionals produce a paint fog that can be difficult to control.

Popular painting brushes.

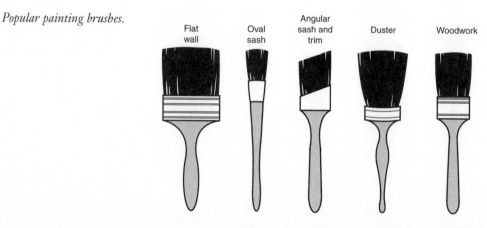

For most do-it-yourselfers, rollers are preferred for large surfaces and brushes for trim and edging. Power rollers are available that siphon paint from the can for application by the roller.

Here are some tips from professional interior painters:

◆ Use the best quality tapered brushes you can afford.

◆ Apply a primer over drywall before using oil-based paints.

◆ Prepare surfaces for painting by cleaning or sanding for a smooth surface.

◆ Read the instructions on the paint can for best results, and to know about safety issues.

◆ Use canvas or paper drop cloths to cover flooring before painting. (Plastic cloths can be slippery.)

◆ Allow the paint to dry before applying additional coats, if needed.

◆ Apply paint evenly, brushing away runs as soon as possible.

◆ Wear old clothes, and paper shoe and head covers (available from paint stores).

◆ Clean brushes if you stop using them for more than two hours. Check the paint can for specific instructions.

Adding Lighting

Sure, you could simply hang a light bulb from the ceiling by a couple of electrical wires and be done with it. But it wouldn't be safe, efficient, or attractive. What's needed here is a lighting fixture. With a little instruction from an electrician, you can install many or all of your new house's lighting fixtures. You can at least learn how it's done for the day when you want to replace a fixture yourself.

Some owner-builders wait until the last minute to purchase and install lighting fixtures. That's okay, as long as there's a budget for them. Once your house is being finished, you can visit local lighting stores and building material retailers to make the final choices. Of course, you probably knew what *type* of fixture you'd install before you started building.

On the Level

The three primary types of lights are incandescent, fluorescent, and quartz halogen. There are also hybrids that make lighting more efficient. Incandescent or filament bulbs have been popular for a century, but they aren't as energy efficient as newer types. Fluorescent tubes contain a gas that helps them be more efficient and reduce energy costs, although they are more expensive to buy. Quartz halogens give off the greatest light for the electricity used, but they are also the most expensive to buy.

Here are some types of lighting fixtures:

◆ Surface-mounted

◆ Shallow shielded

◆ Deep shielded

◆ Rotating

◆ Spot

◆ Closed globe

◆ Lantern

- ◆ Ornate lantern
- ◆ Square
- ◆ Wall
- ◆ Valance
- ◆ Soffit
- ◆ Recessed

There are many more types available. They all serve a function as well as being decorative. Which type you plan for a specific location in your home depends on the area's function. A home office will require different lighting types and fixtures than a bedroom, for example. Your design may vary, but here are some designs for typical room lighting:

- ◆ Living room lighting is reflected off walls and ceilings, except for spot lights for reading areas.
- ◆ Dining room lighting is direct above the table and reflected from recessed wall lights.
- ◆ Kitchen lighting is directed toward working surfaces.
- ◆ Bedroom lighting is diffused except for spot lights for reading.
- ◆ Bathroom lighting is direct near mirrors and diffused elsewhere.
- ◆ Hall and entrance lighting is directed toward steps and reflected off walls.

Lighting fixtures are relatively easy to install. In fact, most fixtures include installation instructions. For example, a two-wire lighting fixture is connected to two wires protruding from the ceiling terminal box, white-to-white and black-to-black. Remove ½" of insulation from the end of the wires, and then twist the appropriate wires together using wire nuts.

Installing a ceiling fixture.

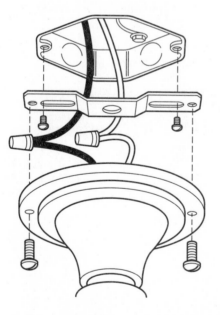

The only tool you may need is a wire stripper. It looks like a pair of pliers with grooves on the inside edge of the cutter. Place the end of the wire in one groove and close the cutter. It removes the insulation around the wire without damaging the wire. Numbers near the grooves tell you the gauge of wire (12, 10, 8) the notches will strip.

Some fixtures, such as ceiling fan-light combinations, have additional wires. Follow the instructions with the unit or hire an electrician to install it. You attach ceiling fixtures to the terminal box using a strap, screws, and nipple that come with the fixture.

In lighting, lumens are more important than electrical wattage. A lumen is a measurement of light output. So selecting lighting fixtures starts with deciding how much light is needed in a specific room or location. Then you choose the appropriate bulb. Finally, you select a decorative lighting fixture for the bulb type and output.

> **CAUTION**
>
> **Code Red** _____
>
> Make sure that the circuits can handle the power needed for specific lighting. And be sure that the fixture is rated at or higher than the bulb's rating. Don't install 75-watt bulbs in 60-watt maximum fixtures. The wires and other components will get hot and may cause a fire!

Installing Appliances

Most appliances are relatively easy to install—just plug them in. The largest appliance in some homes is the electric range and/or oven. That's because they are typically 220v and need up to 50 amps of electrical service. An electric range/oven is connected using a wired three-prong plug or a pigtail, which has a plug at one end and three wire terminals at the other. The terminals are attached to a connection block on the back of or inside the appliance. The three wires are 1) black, 2) red, and 3) green or white.

You'll see 220v marked on some appliances and 250v on receptacles. What's that about? 220v is the average of actual voltage delivered to an appliance, while 250v is the maximum voltage that the receptacle can safely handle. So, a 220v electric stove is plugged into a 250v receptacle.

Electric ranges suck up the energy. Depending on the model, it could be 5-15Kw. Wow! It takes some heavy-duty wire to keep the juice flowing without getting too hot. So, make sure you use the correct plug and wire for your electrical appliance.

Electric dryers don't use as much electricity, but most are also 220v. They use similar plugs and receptacles except that they are typically rated at no more than 30 amps. Check your electrical plan to be sure.

Electric water heaters are often 220v appliances. Follow the manufacturer's suggestions and the above guidelines for installation. Simultaneously, you may want to install a 250v receptacle in your garage just in case Santa brings you an arc welder!

You may also need to install a microwave, dishwasher, trash compactor, garbage disposal, air conditioner, and other built-in appliances. These are all typically 110v appliances with a nearby electric receptacle for plugging in. If possible, give each of these appliances a dedicated circuit. Follow the manufacturer's instructions for installation. Just make sure that the circuit is strong enough to handle the power load by checking the value on its circuit breaker (as described in Chapter 22).

> **On the Level** _____
>
> How can you tell a 30-amp from a 50-amp 250v receptacle? The ground prong on a 50-amp plug is L-shaped; the 30-amp ground prong is straight.

Selecting Furniture

Which furniture should you buy for your new home? If you still have a few dollars in your construction budget, it makes sense to select furniture that matches the design and personality of your new home.

However, most folks are so strapped with debt by this point that they opt to continue using at least most of the furniture they had. If you can't afford that complete dining room set or new sofa and loveseat right now, make a one- or two-year plan for acquiring and replacing furniture. Whatever you do, don't include the cost of 10-year furniture in your 30-year mortgage. Start with what's most needed. It will also give you a chance to watch for sales on your preferences. Many furniture stores also offer deals such as one year with no payment and no interest. If you choose this route, make sure you can pay the full amount when it's due, because the interest accumulates from the date of purchase!

Adding the Finishing Touches

It's really amazing how much difference the finishing touches make. Like what?

- Towel bars, tub and shower grab bars, cabinets, and other bathroom accessories
- Hall and bedroom door mirrors
- Ornate door hardware
- Decorative thresholds
- Built-in book and display shelves
- Plant hangers
- Exterior and interior gingerbread trim
- Wall sconces
- Curtain and drapery hardware such as decorative rods and tie-backs
- Valances, curtains, drapes, and other window treatments
- Entryway coat racks
- Kitchen and bath undercounter storage racks
- Recycle center bins
- Pictures, photos, tapestries, and other wall decorations

Decorating Resources

There's a seemingly endless list of resources to help you decorate your new house. Besides what's included in Appendix B, here are some ideas:

- Lighting shops
- Interior decorators and designers
- Furniture stores
- Paint and decorating stores
- Major appliance stores
- Decorating magazines
- Homeowner how-to magazines
- Specialty home decorating catalogs
- Plan service catalogs
- Your architect
- Friends' and neighbors' homes
- Cable and satellite TV home decorating shows

This is your opportunity to turn a house into *your home!* Gather decorating ideas from all sources, but use only those that fit your tastes and budget. Don't be afraid to experiment and have fun with colors and decor. It's not permanent.

What's next? Moving into your new home!

The Least You Need to Know

- Decorating professionals can give your home a personality of its own—or you can do it yourself.
- Many owners opt to do their own painting to save money for other things.
- Lighting adds both function and decorative value to your new home.
- Finishing touches include adding furniture, appliances, and a few extra features.

Chapter 25

Ready for Occupancy

In This Chapter

- ◆ Managing inspections
- ◆ Cleaning up the building site
- ◆ Getting the occupancy permit and turning on utilities
- ◆ Final inspection: the walk-through
- ◆ Tips for hassle-free moving
- ◆ Resources for preparing for occupancy

This is going to be a shorter chapter because I know you're anxious to get moved in. You've been working hard on building or supervising construction of your home, and the fruit of all that labor is about to pay off.

So what's left to do? You'll need the building department's final inspection and occupancy permit. You'll need to turn on or switch over utilities. You'll want to start getting ready to—yeah!—move in. There's just a little further to go.

Getting Inspected

As your house was being built, numerous inspections were taking place, as I mentioned in Chapters 10 and 11. The specifics of these inspections will be covered here. Exactly when they occur depends on your lender and on the local building department.

As a reminder, a construction loan is a short-term loan to finance the construction of a house. You may build all of the house yourself, participate as the general contractor or a subcontractor, or simply be a knowledgeable owner who will keep everyone honest.

The construction loan won't be paid out all at once. The lender will want to make sure that the construction is progressing as planned. So the lender (with the contractor) will establish milestones where specific tasks are done and can be paid. The payments are called draws because money is drawn out of the construction account.

Here's a common draw schedule:

1. Twenty percent when footings, foundation, first-floor joists, and subfloor (or slab and plumbing rough-ins) are completed

2. Twenty percent when roof and interior are sheathed and all interior partitions are roughed-in

3. Twenty percent when plumbing, electrical, HVAC, insulation, drywall, siding, windows, doors, and roofing are installed

4. Twenty percent when interior and exterior trim, stairs, casework, and countertops are installed

5. Twenty percent when fixtures, plumbing, electrical, lighting, flooring, appliances, driveway, and landscaping is installed and all interior and exterior surfaces are painted

Code Red

Remember: Subsequent construction cannot continue until inspections are passed.

Your lender will either come out to the building site (when called) or send someone official out. Alternatively, some lenders will wait to receive the inspection report from the building department. In any case, make sure the lender knows where you are in the building process and when you will require a draw (as detailed in Chapter 10). Remember, draws aren't paid until construction stages are approved.

The local building department that approved your building plans will also show up (again, when called) to inspect construction at various stages. Here are typical inspection points for a single-family residence construction:

◆ **Foundation inspection** after footing excavations are done, forms are built, and rebar is in place (depending on the type of foundation).

◆ **Slab or underfloor inspection** after the concrete slab, if any, or the subfloor is installed.

◆ **Frame inspection** after all framing, bracing, and roof sheathing are installed and the required utility rough-ins are done.

◆ **Wall inspection** after exterior sheathing and drywall or lathing is installed, but before plaster or tape is installed.

◆ **Final inspection** after the building is completed, the lot finish graded, and the house is ready for occupancy.

◆ **Other inspections** may be required by the building department depending on the type of construction and the plan's complexity; the building department will notify you in advance of additional inspections.

Of course, if there's a specific problem with the frame inspection, it will hold up installing drywall, but your inspector will probably let you go ahead with roofing. Ask. There may be other inspections depending on what you're building and where:

◆ Manufactured homes may require a state building inspector if any modifications are made or appendages are added to the units after they leave the factory.

◆ Installing a well will require a domestic water well permit and inspections.

◆ Installing a septic system will require a septic tank permit and inspections.

◆ In some areas, special plan (such as coastal) permits and inspections are required.

◆ Utility services may require special permits and inspections depending on local building codes.

◆ Installing underground tanks will require a special permit and inspections.

◆ Houses built in the woods may require special forestry permits for clearing land, altering a streambed, or other changes to the natural habitat.

◆ Other entities, such as the local fire protection agency, may require inspections before an occupancy permit is issued.

Remember to find out what permits and inspections you need *before* you begin construction (see Chapter 11).

Finally, if you're using a general contractor, he or she may want you to sign a paper that indicates substantial completion, especially if a draw payment depends on it. Substantial completion is the point at which the contractor feels that he or she has essentially finished the project, but before the final inspection. Don't sign it until you're satisfied that the construction job is virtually done.

Cleaning Up the Place

Ultimately, you are responsible for cleaning up the building site during and after construction. You can hire someone to do it, but if they don't, you must.

Site cleanup depends on many factors. Most important is local building requirements. Some local building codes won't let you store paints and finishes on site. Others will say that scrap materials are a hazard to workers and neighbors that must be immediately cleaned up.

What to do? First, find out what you can and cannot do. Then, if you haven't previously done so, rent a dumpster for on-site trash. Or make sure that contractors clean up after themselves.

The two biggest construction waste categories are lumber and drywall. Fortunately, lumber is easily recycled. Let your crew know where they should place recycled lumber and that it is available to anyone for free. It will quickly disappear. (Just make sure that nothing else disappears with it!)

Drywall is more of a problem. In fact, drywall takes up so much space in many landfills that local codes require recycling it. Fortunately, your building materials supplier or drywall contractor can help you find ways of recycling it. Also contact local recycle centers for their recommendations.

Code Red

Don't bury trash on the building site. Not only is it illegal in most places, it's also dumb. Lumber scrap can become the new home of termites and other pests. Gypsum and paints can leach into the water table. Concrete can get in the way when installing landscaping. Recycle it instead!

Getting the Occupancy Permit

Before you can *legally* live in your new house, the local building department must make a final inspection for health and safety codes. Only then will it issue the owner an occupancy permit or certificate of occupancy. In addition, any utilities that serve your house are temporary and can't become permanent until the occupancy permit is issued.

So what will it take? The building inspector issues a certificate of occupancy to the owner by the building inspector once the final building inspection approval is granted. Depending on the local building department, this may mean that you can begin occupying the structure once the interior is finished. You don't have to wait until it is landscaped or a driveway is finished. In fact, you may be able to get an occupancy permit before the garage is constructed or before carpeting is installed. Ask your local building department for specifics.

So what? If you are paying rent on a temporary residence, you can get into your new home faster.

Of course, your lender may not release funds until the carpet is down or the garage in. Or they may require that the materials for final projects are purchased, to be installed by a specified date.

In most locations, occupancy permits are required for all residential (and commercial) structures. It doesn't matter whether the residence is a conventional house, a yurt, a log home, a straw-bale house, a manufactured home, or whether you built all of it or none of it yourself. If it required a building permit to start the project, it will probably require an occupancy permit when it's done.

Ordering Utility Services

Time to turn on the power. And connect up the water and sewer. By the time you've finished construction, these services are ready for distribution to your house. Electrical service is wired. Water and sewer are plumbed. They may even be temporarily connected. The utility provider will require a copy of the certificate of occupancy before turning on the service or switching it from temporary to permanent.

In most cases, the only temporary services you can get are electrical, water, and maybe sewer. Gas and other services will wait until you get an occupancy permit. Fortunately, telephone service will only have to wait until you have electrical power (unless it's just for a cell phone). So it's going to be your job to find out what each of the utilities requires for permanent service and to make sure they get it.

Many utility services will make a final inspection of your residence if asked nicely. While there, they can show you how to finish hookups or light gas pilots.

> **Code Red**
>
> Why not just leave services on temporary status and live in the house? First, because it's illegal—meaning there may be fines or worse if you do. Second, because it's expensive—temporary service rates are often higher than permanent rates.

Walking Through One More Time

If you are working with a general contractor and/or a lender, you will want to make your own inspections along the way. The most important is your final inspection or walk-through. It's when everyone involved in this construction project meets to see the final result.

Participants in the final inspection may include you, your partner, your lender, and your primary construction advisor. That may be your GC or a contractor whom you've relied on throughout the building process.

Alternatively, you may decide to make two or three "final" walk-throughs. The one with your contractor/advisor will cover construction issues. The walk-through with your lender will focus on financial issues. The inspection with your partner will be about livability. What should you be looking for?

> **Building Your Vocab**
>
> A **punch list** is a written list of unfinished or incorrectly finished items that must be finished or corrected before the owner will accept the project from the contractor.

- Conformance to the building plans
- Quality of workmanship
- Unfinished jobs
- Substandard materials

By spending more time on design and on selecting the best tradespeople, the final walk-through should be easy and offer no surprises. That's your goal—and the goal of this book. You also want to make sure that the contractors have done their jobs. You can make a *punch list* to give to the contractor.

Preparing to Move In

Congratulations! You're ready to move into your new home. As author of *The Complete Idiot's Guide to Smart Moving* (Alpha Books, 1998), I can offer you some expert tips on moving in:

- Start a Moving Notebook (or use your Home Book) early so you can begin organizing the move to your new house. Include a list of tasks and contacts that you can check off as they're done.
- Once you know the date you'll be moving in, send out change of address cards to family, friends, healthcare providers, and services.

◆ Contact utilities such as the phone company ahead of time to arrange for service before your move-in date.

◆ Decide a month or two in advance whether you plan to pack and move your things or to hire a service to do some or all of it. If you choose to use a moving company, begin shopping early for the best prices. If you'll be moving yourself, arrange for boxes and rolls of tape to be dropped off, and check prices for truck rentals, if necessary.

◆ To help them feel ownership in the move, ask children to pack their most precious belongings and unpack them first when they arrive.

◆ Instruct older children on packing methods to teach them new skills and independence they will need in coming years.

◆ Make sure your insurance company knows about your new house and your moving plans.

◆ Start packing items early that you know you won't use, such as books and collectibles. Dust them as you pack so when you unpack them they'll be ready to put on the shelves.

◆ Label all boxes according to what room they will go in. If the contents of a box are fragile, make sure it's clearly labeled on the box.

◆ If possible, set aside a room or the garage to store packed boxes and large unused items in preparation for moving.

◆ Carefully plan out the two most important days: moving-out day and moving-in day.

◆ Set aside a special "open me first" box with cleaning supplies, sheets, towels, toilet paper, paper towels, pet supplies, toiletries, and other necessities that you'll need right away. Make sure it's clearly labeled so you can find it easily.

◆ Make sure you take some time to have fun and relax to reduce the stress of moving.

◆ Contact the Internal Revenue Service (www.irs.gov) for Publication 521 on deductible moving expenses.

Occupancy Resources

Your primary resources in preparing for occupancy are your …

◆ Local building department
◆ Building inspector(s)
◆ Lender
◆ Contractor(s)
◆ Local utility services
◆ Mover or moving equipment suppliers

Probably the most important resource is your Home Book. It has been your schedule, journal, notepad, contact list, and reminder since beginning this fascinating project. It can serve you well now.

Your new home is more than just a structure. It's landscaping, driveways and walkways, fences and decks, and other outdoor living components. They're coming up in the next and final chapter.

The Least You Need to Know

◆ Make sure you're ready for the various building and lender inspections so that you can get the occupancy permit without hassle.

◆ You are responsible for cleaning up the building site during and after construction.

◆ Remember to switch temporary utility services to permanent once the certificate of occupancy is issued.

◆ Take time to walk through your new home, making notes of any last-minute fixes needed before or after you move in for good.

◆ Make your moving-in experience enjoyable and hassle-free by planning ahead.

Chapter 26

Outdoor Options

In This Chapter

- ◆ Planning your landscaping
- ◆ Selecting the best materials
- ◆ Installing fences, decks, pools, and outbuildings
- ◆ Installing paths, walks, and driveways
- ◆ Resources for outdoor options

Congratulations! You've moved into the house that you designed and participated in building. Now it's time to make it into more of a home.

What will that take? Landscaping, fences, decks, pools, sunrooms, outbuildings, paths, walks, driveways, and other amenities. You have lots of choices. This final chapter covers these and related topics, and offers additional resources for making your house a home.

Planning Your Home's Landscaping

Among all the scrolls of construction plans, you may have one that's labeled "landscape plan." You can start planting and building things without a plan, but it will be somewhat like building a house without plans: haphazard. So even if you sketch out your best ideas in your Home Book, make sure you have a good landscape plan in place before starting.

Here are four things to think about when planning landscaping:

- ◆ Nature
- ◆ Harmony
- ◆ Concealment
- ◆ Maintenance

First, look at the natural elements on your building site to decide what effect they will have. Will the trees, fence, or shrubs change the amount of sunlight that gets into your yard? Will they increase the shade? Will they redirect or stop the wind? If so, how much and during what part of the year? Will a planned tree drop leaves into the pool? Will a natural element cast too much shade into your garden area?

Next, consider harmony. The main idea of landscaping is harmony of purpose and of design. Will the ground cover blend in with your stone wall or will it detract from it? Are the colors of nearby flowers complementary to your home and other structures?

Concealment applies to landscaping elements that cover something else: a view, a fence, a structure. Ivy can be grown to conceal a fence. Trees can be planted to conceal a view. Shrubs can be planted to conceal an air conditioner.

Finally, consider maintenance. Maintenance is important because whether the landscape element is a flower bed, ground cover, or a fence, it should not require undue maintenance—unless you're looking for another pastime. Some variations of ivy can take over a yard in a couple of seasons. Other plants can thrive on neglect. How much time do you want to spend maintaining your landscaping? Will it be a hobby or a chore? Of course, there are other landscaping considerations such as soil, water, drainage, and local growing conditions.

 Ka-ching!

If you don't yet have a landscape plan, first draw a map of your building site to scale. You can use a copy of the plot plan that you filed with the local building department. Then decide what natural elements you want to keep, how you can harmonize the house, how you can best conceal, and how much maintenance you want to do to keep up your yard. Look to your neighbors' yards for ideas on what to do and not do with your new yard.

Selecting Outdoor Materials

Outdoor design elements include living things like trees and plants as well as inanimate things like fences and decks. Plants are plants, but fences are made up of component materials.

The most popular outdoor material is wood in the form of lumber. Because lumber is exposed to sun, moisture, fungi, and crawly things, it should be treated with a preservative. You can do it yourself or you can buy wood that is already treated, called pressure-treated lumber. Some woods such as redwood and cedar have natural preservative qualities, though you may want to add some protective finish to maintain these qualities.

Other outdoor materials include cement, cement blocks, bricks, and other masonry components for walks, walls, driveways, and decorative structures.

You can buy these and other outdoor materials through your primary building materials supplier. Or you can get them at landscape stores along with the plants and trees for your yard.

Landscaping Your New Home

Your building and site plans included specifics on the final grade of the home's site. So the first step in landscaping is to make sure that the foundation is backfilled and that the site is graded to the plans. Also, contour the soil for best drainage and to make walking easier.

If your landscape plans include a sprinkler system, now's the time to install it. Get it in once the site is graded and before you place plants. What you plant depends on why you plant. That is, you may decide on a specific type of tree because it is low maintenance while quickly growing to a height that will block an unsightly view.

Two important factors when selecting trees and large plants include growth rate and root system. You want trees that will grow to a specific height and shape and then slow down so that they don't overtake everything around them. It's the same issue underground; you don't want a root system that stifles other plant growth, clogs drains, or lifts nearby pathways.

Even if you have loads of good books about landscaping, ask for localized advice from area nurseries. Plants that may thrive just a few miles from your house may not do as well where you live because of soils, drainage, or microclimates.

Landscaping usually means adding a lawn. Seed or sod? For immediate coverage, sod lawns are the best choice, although they are more expensive than seed lawns. They are also preferred for new building sites where the topsoil has been removed. In many locations, it's important to get the lawn in before fall to minimize winter runoff. That means installing a sod lawn over a few days rather than the months required to establish a seed lawn.

Building Good Fences

No one knows when the first fence was built, but the largest is still at least partially in place. The Great Wall of China was begun 2,300 years ago, and many sections still stand.

Of course, your building lot is smaller than China. But the purpose of fencing is the same: to keep out marauding hordes. Actually, people install fences for many reasons:

On the Level

Dimensional lumber used for decks, fences, and other outdoor structures is available through building material suppliers. Dimensional lumber is sized similarly to housing lumber such as 2" × 4" nominal. In addition, boards of 1" × 4" through 1" × 12" are popular for decking and fencing.

- Mark property boundaries
- Keep people and animals out
- Keep people and animals in
- Privacy
- Control soil and erosion
- Reduce or control noise
- Control landscaping
- Decorative or aesthetic purposes

Fences are built from lumber, railroad ties, plywood, hardboard, aluminum, iron, steel, logs, wire, brick, stone, poured concrete, split wood, bamboo, canvas, fiberglass, adobe, and just about anything else you can think of.

The most common types of residential fences are rail, picket, board, chain-link, and masonry. Which style you choose depends on what you're trying to accomplish, as well as possible neighborhood regulations. To keep pets in or out, chain-link fencing is preferred, although it's not the most appealing aesthetically. For a rustic backdrop to shrubs, board fences work well. For a homey look, white picket fences are a good choice. Check with your local building department and any neighborhood association for requirements on fences, decks, pools, and other yard structures. In some communities, fence heights are limited in front yards and especially on corner lots.

Code Red

Be aware that local building codes may or may not require a permit before you construct a fence or deck. Or it may depend on the fence's height. Check with your building department for requirements.

Wooden fences are made up of vertical posts, horizontal stringers, and horizontal or vertical siding. Chain-link fences have posts, a top rail, and galvanized fabric stretched between the posts. Masonry fences are built of brick and/or block, and mortar.

Common board fence designs.

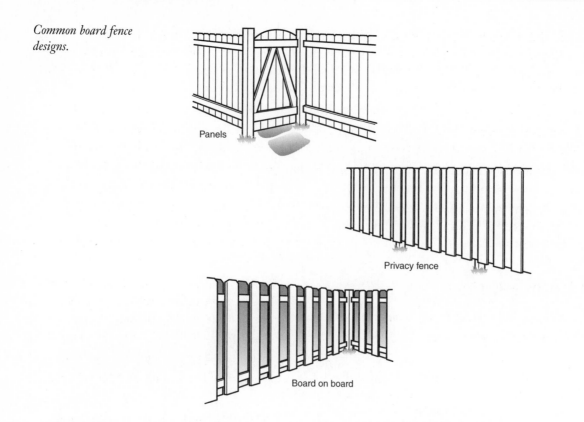

Panels

Privacy fence

Board on board

Unless you enjoy jumping over fences, you'll need to install gates. Their function is obvious: to let you walk through the fence. So designing a gate is easy. Make sure it is wide enough for you and anything else (lawn tractor, trailer) that must go through the fence. Make the gate look like it is part of the fence. And make sure that it is strong enough to withstand long-term use.

Expanding Your Home with a Deck

Decks are valuable outdoor additions to a home and can greatly enhance its market value. They can be outdoor rooms, or they can be yards for otherwise unusable lots. Our home includes a partially covered deck that is half the square-footage of our house, expanding our living area throughout the year.

Most decks consist of six parts: footings, posts, beams, joists, decking, and railings. Sound familiar? These components serve the same functions in decks as they do in houses, except that decking is the floor, and railings or balusters are the walls.

Deck components.

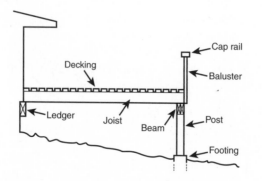

Cap rail

Decking

Baluster

Ledger

Joist

Beam

Post

Footing

Building a deck is similar to building a house, only easier. You can hire someone to build the deck for you, or you can do it yourself. The first step in building your deck is to mark off the deck area using string and batter boards. Make sure that it is level and square. The string will help you visualize the size and appearance of the finished deck, and will also serve as a guide for excavation and post placement.

The second step is to prepare the site. Clear the area of soil or sod to a depth of two to three inches and about 2 feet wider than the proposed deck. Then locate and dig holes for footings. In many areas, that means digging a hole at least 24 inches deep (depending on the local frost depth) and filling it with wet concrete. On top of the dry footing goes a concrete block or the post, so, as necessary, install a fastener in the wet concrete.

Beams are horizontal members attached to each of the posts. If the deck is attached to the house, install a header beam on the house. Perpendicular to the beams, install the joists, spacing them 16", 24", or 32", depending on the joist size (2" × 6", 2" × 8", 2" × 10", respectively). These are general guidelines; make sure you're following local building code.

The decking is installed next, spaced ¼" to ½" apart to allow for expansion and water runoff. Depending on the deck's design, you'll build the railings on top of the decking (like platform construction) or as extensions of the post frame. You can then add top and side rails for function and decoration. Additionally, you can build benches or tables, attaching them to the post and rails. You can also build a trellis for shading the deck.

Finally, design and install stairs (see Chapter 19) as needed to access the deck. Decks are like houses in that they can be as simple or as complex as you desire. Your deck can be an extension of an adjacent room—or it can be your summer home.

Sunrooms are room extensions or walk-in windowed rooms. They distribute sunlight and air circulation to the home. You can build them over an extension of the home's foundation or over an attached deck. Sunrooms are available as do-it-yourself kits or as finished rooms built by a specialized contractor.

Installing a Pool

If your house plan calls for a swimming pool, congratulations! They can be valuable additions to the livability of your home (although they typically are not good financial investments). Maybe you've saved enough by building your own home to consider installing a swimming pool yourself or having one installed for you!

There are two popular types of pools: in-ground and above-ground. Costs vary dramatically based on size, shape, materials, and the difficulty of construction. The least expensive in-ground pool, the vinyl-lined pool, is built with vinyl sides supported by an aluminum, steel, plastic, masonry, or wood frame. At the other end of the price spectrum is a concrete swimming pool. Concrete? Well, air-sprayed concrete or Gunite, to be precise. Construction requires excavating a big hole, lining it with reinforcement bar, and spraying it with a special concrete. A waterproof plaster is then hand-troweled over the concrete for a smooth surface.

In between vinyl and concrete is the fiberglass pool. These are either built in a factory and trucked to the site or built on-site. Factory-built fiberglass pools are typically cheaper.

Above-ground pools require little or no excavation, so they are less expensive to install. They come in all shapes and sizes, from oval to rectangular and from 10' round to 16' × 32'.

Your new swimming pool can be part of or next to a deck attached to your house. Or it can be away from the house. What should you look for in a swimming pool?

- Lap swimmers need a surface area of at least 4' × 6'.
- Divers need a depth of at least 7½'.
- Allow 36 s.f. (6' × 6') of water surface for each swimmer.
- Allow 100 s.f. (10' × 10') of water surface for each diver.

◆ Young children should have a shallow pool or portion of 3' to 4' depth.

◆ Add a deck to your pool for the same reason you add one to your house: to extend functionality.

◆ Extend your pool season with a pool cover.

What equipment will your pool require? That depends on the local climate, how much you use the pool, and how you use it. A weekends-only pool in San Diego may not need a heater. An active family's pool in Minneapolis may need a year-round cover and a heavy-duty heater. Both will need a water filtration and chlorination system.

Typical in-ground pool heating and filtration system.

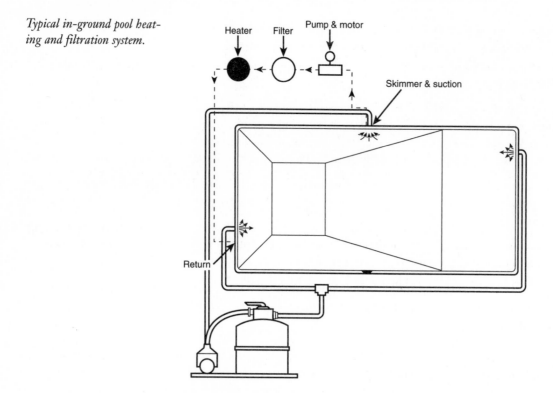

Hot tubs and spas are increasingly popular. Fortunately, most are relatively easy to install and can be heated using 110v electricity. Others are heated by a wood stove or solar panels.

Other Buildings

Your landscaping can include outbuildings such as a gardening shed, storage building, unattached garage, workshop, hobby shop, getaway office, or even a mother-in-law house.

On the Level

Consider adding a greenhouse to your landscaping. They are relatively easy to construct and you can use them for growing flowers or vegetables year-round. Many greenhouse kits are available for less than $1,000.

Constructing an outbuilding is similar to building your home, just easier. If it won't be large or habitable, a simple slab or even a wooden foundation may be sufficient. Shed kits and plans are available through larger building material stores. They include materials and/or plans for construction. Sizes are usually in 4' increments starting at 8' × 8' sheds and going up to 32' × 40' pole buildings.

Some outbuildings are built just for fun. A gazebo is a freestanding structure for relaxation or entertainment. The roof can be solid or lattice. Sides are typically open. An aerie is a small gazebo with benches for sitting. Sunshades, also known as trellises, are even more open, with a lattice top and no walls.

Installing the Driveway

How are you going to get from the road to your garage? A driveway, of course. But what kind of driveway? The decision is made for you if you're building a standard home on a standard subdivision lot. The street curbs are already in, and there's not enough room to have anything but a straight driveway from there.

If you have a large lot or acreage, planning a driveway can be more complicated. It may need to go up or down a slight slope, at an angle, or even across some marshy land. More challenges. Fortunately, you've planned this out already and know approximately where the driveway is going.

On the Level

What's the best size for your driveway? For a two-car garage, you'll want it to be as wide as two cars in front of the garage, but you won't have two lanes of traffic coming in from the road. So, it will start narrower at the road and widen near the garage. Have or plan to buy a recreational vehicle? Better leave a level parking pad for it. Plan to fill up the garage with stuff and park outside? Make sure the driveway in front of the garage is large enough so you don't have to park on the lawn.

What hard surface will you use for your driveway? Asphalt? Concrete? Rock? Concrete is the most durable, and the most expensive. Asphalt or macadam is less expensive (in some areas), but typically requires recoating every five to ten years depending on the climate. Gravel rock is cheaper, but ruts easily and is only recommended for long driveways where cars don't sit still long.

Some builders install a rock driveway during house construction, and then smooth it out and pour a cement drive once construction is done. This saves wear and tear from heavy trucks. What works best for your location depends on local climate and soil.

Making Walks and Paths

Who installs the sidewalk in front of your home (if there is one)? Good question. You probably got the correct answer as you estimated costs and filed for building permits. In some locations, the sidewalk is city property and is maintained by the local road department. In other places, the sidewalk is yours to pour and maintain. In many subdivisions, it's yours but the property developer already installed it and added it to the cost of your lot.

In any case, you will be responsible for designing and installing any walks and paths on your property. That means the walk from the garage or driveway to your front door, any walks that lead to the side or backyard, and any walks or paths in the backyard. Fortunately, you have many options:

◆ Rock or gravel (small rock) contained by 1" × 2" boards

◆ Concrete, stone, or wooden steps

◆ Bark chips

◆ Poured concrete

◆ Asphalt

◆ Wooden path

◆ Dirt

You can construct walks and paths by first laying out where it will go, establishing its width boundaries, and preparing the soil. Some soils only need removal of the top two inches before laying down rock, gravel, bark, or even concrete. Others require more preparation. Concrete won't require reinforcement bar unless it is thick or heavy objects will traverse it.

Outdoor Resources

There are many valuable resources for landscaping your home. Among them are …

- Popular home and landscape magazines
- Local landscape architects
- Landscape nurseries
- College courses on landscape design and maintenance
- Homes and homeowners in your new neighborhood
- Additional resources in Appendix B

Congratulations again! You've learned how to build your own home, saving many thousands of dollars. The money probably went right back into your home, giving you more for every dollar spent. Thank you for sharing these pages with me. I sincerely hope you enjoy your new home!

The Least You Need to Know

- Planning and installing your own landscaping can not only save you money, but also give you additional pride of ownership.
- Fences are functional and decorative additions that add value to your home.
- If you've saved enough money, consider installing a swimming pool or other amenity.
- How your driveway and walks are installed depends on the size of the building lot, the soil, local climate, uses, and related factors.

Home Construction Glossary

ABS (acrylonitrile-butadine-styrene)   Material used for rigid black plastic pipe in DWV systems.

acceleration clause   An option in a mortgage agreement that says the unpaid balance must be paid immediately if payments are missed or if terms of the agreement are not met.

access hole   An opening in a ceiling or floor that provides access to the attic or crawl space.

adjustable-rate mortgage (ARM)   A mortgage in which the rates of interest and payment change periodically, based on a standard rate index. In most cases, the ARM has a cap or limit on how much the interest rate may increase.

air-dried lumber   Lumber that has been dried naturally by air, with a minimum moisture content of 15 to 20 percent.

amortization   The building of equity and reduction of debt through regular monthly payments of principal and interest.

anchor bolt   A bolt used to secure a wooden sill plate to concrete or masonry patio or foundation.

APR   Annual percentage rate. *See* principal and interest.

apron   The flat piece of inside trim that is placed against the wall directly under the stool of a window.

areaway   An open subsurface space adjacent to a building used to admit light or air or as a means of access to a basement.

armored cable   A flexible metal-sheathed cable used for indoor wiring. Commonly called BX cable.

asphalt   A black bituminous material used to pave and waterproof roofing and other building materials. Usually a residue of evaporated petroleum.

assumable mortgage   A mortgage in which the buyer takes over the payments on the seller's old mortgage and pays the seller the difference between the selling price and the remaining loan balance.

attic ventilator   The openings in the roof or gable; allows for air circulation. Sometimes the power-driven fans that force air into the attic.

backfill   To replace excavated earth into a trench around and against a basement foundation.

backhoe   A machine that digs deep, narrow trenches for foundations and drains.

ballast   A magnetic coil that adjusts current through a fluorescent lighting fixture.

balloon mortgage   A real estate loan in which some portion of the debt remains unpaid at the end of the term, resulting in a single, large payment due at the term's end.

baluster   A vertical member of the railing of a stairway, deck, balcony, or porch.

band joist   A joist nailed across the ends of the floor or ceiling joists. Also called a *rim joist*.

base molding   A strip of wood used to trim the upper edge of a baseboard.

base shoe   A strip of wood used to trim the bottom edge of the baseboard, next to the floor. Also called a carpet strip.

baseboard   A trim board placed at the base of the wall, next to the floor.

batten   Narrow strips of wood used to cover joints or as decorative vertical members over wide boards, such as on fences.

batter boards   Horizontal boards nailed to posts or stakes to mark the corners or ends and the proper level of foundation walls or rows of piers.

bay window   Any window—square, round, or polygonal—projecting outward from the wall of a structure.

beam   A structural member, usually steel or heavy timber, used to support floor or ceiling joists or rafters.

bearing wall   A wall that supports any vertical load in addition to its own weight.

bedding   A layer of mortar into which brick, stone, or tile is set.

bevel siding   A type of exterior siding that is made by sawing square-surfaced boards diagonally to produce two wedge-shaped pieces of siding. Also called clapboard siding.

bird's-mouth   The triangular cut in a rafter that rests on the top plate of a stud wall. Consists of seat (horizontal) cut and heel (vertical) cut.

blind nailing   Nailing in such a way that the nail heads are not visible on the face of the work.

blue lines or **blue prints**   Reproductions of original construction documents that produce blue lines on a white background.

board foot   A unit of lumber equal to a piece one foot square and one inch thick; 144 cubic inches of wood.

board lumber   Yard lumber less than two inches thick and two or more inches wide.

box   A metal or plastic container for electrical connections. More correctly called a *junction box*.

brace   A piece of lumber or metal attached to the framing of a structure diagonally at an angle less than 90 degrees, providing stiffness or support.

branch   In a plumbing system, any part of the supply pipes connected to a fixture.

bridge loan   A loan that allows financing to cover mortgage payments for six months to a year so that you can sell your old house.

bridging   Narrow wood or metal members placed on the diagonal between joists. Braces the joists and spreads the weight load.

build-up roof   A roofing system composed of alternative layers of roofing felt and tar that are topped with gravel or crushed rock. Usually used on flat or slightly sloped roofs.

builder set   A set of the minimum required construction drawings to get a building permit.

builder's paper   Usually, asphalt-impregnated paper or felt used in wall and roof construction; prevents the passage of air and moisture. Also known as building paper, and typically available in 15-lb. and 30-lb. densities.

building code   The collection of legal requirements for the construction of buildings.

building drain   The lowest horizontal drainpipe in a structure. Carries all waste out to the sewer.

butt joint   The junction where the ends of two timbers or other members meet in a square-cut joint.

cable   An electricity conductor made up of two or more wires contained in an overall covering.

cantilever   A construction system that allows a horizontal structural member to project beyond its support.

cap plate   The framing member nailed to the top plates of stud walls to connect and align them. The upper most of the two top plates, sometimes called the double-top plate.

carriage   In a stairway, the supporting member to which the treads and risers are fastened. Also called a *stringer*.

casement   A window sash on hinges attached to the sides of a window frame. Such windows are called casement windows.

casing   Moldings of various widths and forms; used to trim door and window openings between the jambs and the walls.

caulk   Viscous material used to seal joints and make them water- and airtight.

caulk gun   A tool used to apply caulk.

check valve   A valve that lets water flow in only one direction in a pipe system.

circuit   The path of electric current as it travels from the source to the appliance or fixture and back to the source.

circuit breaker   A safety device used to interrupt the flow of power when the electricity exceeds a predetermined amount. Unlike a fuse, you can reset a circuit breaker.

clapboard siding   A type of exterior siding that is made by sawing square-surfaced boards diagonally to form two wedge-shaped pieces. Also called *bevel siding*.

cleanout   An easy-to-reach and easy-to-open place in a DWV system where obstructions can be removed or a snake inserted.

collar beam   A beam connecting opposite pairs of rafters. Also called a tie beam.

column   A vertical support—square, rectangular, or cylindrical—for a part of the structure above it.

common rafter   One of the parallel rafters, all the same length, that connect the eaves to the ridge board.

concrete   A mixture of aggregates and cement that hardens to a stonelike form and is used for foundations, paving, and many other construction purposes.

conductor   Any low-resistance material, such as copper wire, through which electricity flows easily.

conduit   A metal, fiber, or plastic pipe or tube used to enclose electric wires or cables.

construction documents   The complete set of drawings and written specifications for the construction of a building, being a part of the legal contract for the construction.

construction loan   The short-term loan that pays for the land, materials, and labor required to construct a new building. This loan is funded at the commencing of construction and is usually paid off at the end of the construction period by the permanent mortgage loan.

contract of sale   An agreement that binds the seller and buyer to a price for the home.

conventional mortgage   An agreement between a buyer and a seller with no outside backing such as government insurance or guarantee.

convertible ARM   An adjustable-rate mortgage that can eventually be converted to a fixed-rate mortgage.

coped joint   A junction in which one end of a piece of molding is cut to fit the shaped face of another. Provides a tight fit and an alternative to mitering.

corner bead   A strip of formed sheet metal or wood that protects the corner of a stucco or plaster wall.

corner boards   Boards used to trim the external corners of a structure.

corner braces   Diagonal braces at the corners of a structure that stiffen and strengthen the walls or post-to-beam connections.

cornice   The overhang of a roof at the eaves, usually consisting of a fascia board, soffit, and decorative molding. Any decorative member placed at or near the top of a wall.

cost per square foot (CPSF)   The figure obtained by dividing the total cost of construction (construction contract price) by the area of livable or air-conditioned square feet in the constructed building. This figure does not include cost of land or spaces that are not air-conditioned, such as garages, porches, terraces and decks, but is most often used in comparing construction prices.

countersinking   Boring the end of a hole for a screw or bolt so that the head can be brought flush with or below the surface. Also, sinking or setting a nail or screw so that the head is flush with or below the surface.

courses   The alternative layers or rows of material, such as shakes or shingles on a roof or bricks in a wall.

cove molding   A molding with a concave face. Usually used to trim or finish interior corners.

CPVC (chlorinated polyvinyl chloride)   The rigid white or pastel-colored plastic pipe used for supply lines.

crawl space   A shallow space between the floor joists and the ground; usually enclosed by the foundation wall.

cripple   Any framing member that is shorter than other members in the same structure.

cripple studs   Short studs surrounding a window or between the top plate and end rafter, in a gable end or between the foundation and subfloor. Also called *jack studs*.

cross bridging   Diagonal bracing between adjacent floor joists, placed near the center of the joist span to prevent joists from twisting.

crown molding   A convex molding used horizontally wherever an interior angle is to be covered (usually at the top of a wall, next to the ceiling).

current   The movement or flow of electrons, which provides electric power. The rate of electron flow as measured in amperes.

d   *See* penny.

dado joint   A joint where a dado is cut in one piece of wood to accept the end of another piece.

dead load   The weight of the permanent parts of a structure that must be supported by the other parts of the structure. Does not include the weight of the people, furniture, and other things that occupy the building.

debt-to-income ratio   A ratio used by lenders to decide if an applicant qualifies for a loan; the ratio is the total amount of debt divided by the total gross monthly income.

default   A borrower's failure to comply with mortgage payments and provisions.

density   The mass of substance in a unit volume. When expressed in the metric system, it is numerically equal to the specific gravity of the same substance.

details   Drawings used to clarify complicated construction features.

dimension lumber   Yard lumber from two inches up to, but not including, five inches thick and two or more inches wide. Includes joists, rafters, studs, planks, and small timbers.

direct nailing   Driving nails so that they are perpendicular to the surface or joint of two pieces of wood. Also called *face nailing*.

doorjamb   The case that surrounds a door. Consists of two upright side pieces called side jambs and a top horizontal piece called a head jamb.

dormer   On a sloping roof, a projection with a vertical wall for one or more windows.

downspout   The vertical pipe, usually made of metal, that carries water from the gutters to the ground.

draw request   Monthly request by a contractor to be paid for the materials and labor installed into the project during the previous 30 days, to be drawn from the construction loan.

drip cap   A wood molding or metal piece placed above a door or window frame. Causes water to fall beyond the outside edge of the frame.

drip edge   A piece of angled metal placed along the edge of a roof. Causes water to fall off the outer edge rather than run down the fascia or cornice.

drop siding   Exterior siding, usually ¾-inch thick, shaped to a specific pattern, such as tongue and groove or shiplap.

drywall   Panels consisting of a layer of gypsum plaster covered on both sides with paper, used for finishing interior walls and ceilings. Also called wallboard, gypsum wallboard, and Sheetrock, a trade name.

ducts   Pipes that carry air from a furnace or air conditioner to the living areas of a structure.

due-on-sale clause   An option in a mortgage agreement that states that the loan must be paid in full if the home is sold.

DWV (drain-waste-vent)   An acronym referring to all or part of the plumbing system that carries waste water from fixtures to the sewer and gases to the roof.

earnest money   A deposit to show goodwill on the part of the buyer.

eaves   The overhang of a roof; projects beyond the outside walls.

edge grain   Lumber that has been sawed parallel to the pith of the log and approximately at right angles to the growth rings.

elevations   Representational drawings of interior and exterior walls to show finish features.

equity   That portion of the value of a property left over after deducting the amount still owed on that property.

escrow   An account where a deposit is held until closing; it is then applied to the down payment.

expansion joint   A fiber strip used to separate blocks or units of concrete to prevent cracking due to expansion as a result of temperature changes. Often used on a larger concrete foundation and floor slabs.

face nailing   Driving nails so that they are perpendicular to the surface or joint of two pieces of wood. Also called *direct nailing*.

fascia   The front surface of the cornice or eaves.

fascia board   The flat board, typically perpendicular to the ground, that forms the face of the cornice or eave; the member to which most gutters are attached.

FHA   Federal Housing Administration, which charges below-market interest rates with lower down-payment requirements.

finish carpentry   The fine work—such as that for doors, stairways, and moldings—required to complete a building.

finish electrical work   The installation of the visible parts of the electrical system, such as the fixtures, switches, plugs, and wall plates.

finish plumbing   The installation of the attractive visible parts of a plumbing system such as plumbing fixtures and faucets.

fire-stop   A solid, tight piece of wood or other material to prevent the spread of fire and smoke. In a frame wall, usually a piece of 2×4 cross-blocking between studs.

fitting   In plumbing, any device that connects pipe to pipe or pipe to fixtures.

fixed-rate mortgage (FRM)   A loan with an unchanging interest rate.

fixture   In plumbing, any device that is permanently attached to the water system of a house. In electrical work, any lighting device attached to the surface, recessed into, or hanging from the ceiling or walls.

flagstone   Flat stones, usually from one to three inches thick, used for paving, steps, or walls.

flashing   Sheet metal, roofing felt, or other material used at the junction of surfaces to prevent the entry of water.

floor plan   Representational drawing of everything that constitutes the house.

flue   In a chimney or vent, the opening through which smoke and other gases pass.

fly rafters   The end rafters of a roof overhang, supported by lookouts and the sheathing. Also called rake rafters or barge rafters.

footing   The rectangular concrete base that supports a foundation wall or pier or a retaining wall. Usually wider than the structure it supports.

form   The temporary structure, usually of wood, that gives shape to poured concrete until it is dry.

foundation   The supporting portion of a structure below the first-floor construction or below grade, including the footings.

frame   The enclosing woodwork around doors and windows. Also, the skeleton of a building; lies under the interior and exterior wall coverings and roofing.

freize   The horizontal decorative member of a cornice. Also, any sculptured or decorative ornamental band on a building.

frost line   The depth of frost penetration in the soil. This depth varies in different parts of the country.

furring   Narrow strips of wood attached to walls or ceilings; forms a true surface on which to fasten other materials.

fuse   A safety device for electrical circuits; interrupts the flow of current when it exceeds predetermined limits for a specific time period.

gable   The vertical, triangular part of a structure between the slopes of a roof.

gable end   An end wall that has a gable.

gambrel   A roof with two slope angles; a steep one at the edge of the building and a shallow one at the center.

general conditions   A general listing of the requirements and understandings upon which a construction contract is based.

girder   A large beam of steel or wood; supports parts of the structure above it.

government-insured mortgage (GIM)   A loan in which the government promises to make good on the insured portion of the mortgage if the borrower defaults. Usually, these loans do not require down payments; however, there are strict eligibility requirements.

grade   The ground level surrounding a structure. The natural grade is the original level. The finished grade is the level after the structure is completed.

graduated payment mortgage (GPM)   A loan with an interest rate that starts out low and increases gradually.

grain   The direction, size, arrangement, appearance, or quality of fibers in wood.

groove   A rectangular channel cut with the grain of a piece of lumber.

ground   Connected to the earth or something serving as the earth, such as a cold-water pipe. The ground wire in an electrical circuit is usually bare or has green insulation.

grout   Mortar that can flow into the cavities and joints of any masonry work, especially the filling between tiles and concrete blocks.

growing equity mortgage   A loan in which the monthly payment increases annually, with the increase applied to the principal.

gusset   A flat piece of metal, wood, or plywood; used to support the connections of framing lumber. Most often seen on joints of trusses and the joist splices. Also called a *truss plate*.

gutter   A wood, metal, or plastic channel attached to the eaves of a house; catches and carries rainwater to the downspouts.

gypsum wallboard   Panels consisting of a layer of gypsum plaster covered on both sides with paper, used for finishing interior walls and ceilings. Also called *drywall*, *Sheetrock*, and *wallboard*.

hanger   Any of several types of metal devices for supporting pipes, framing members, or other items. Usually referred to by the items they are designed to support—for example, joist hanger or pipe hanger.

hard costs   All the costs associated with a project that purchase real (hard), resalable components, such as land, building materials, or construction labor.

hardboard   A synthetic wood panel made by chemically converting wood chips to basic fibers and then forming the panels under heat and pressure. Also called Masonite (a brand name).

hardwood   The wood of broadleaf trees, such as maple, oak, and birch. Although hardwood is usually harder than softwood, the term has no actual reference to the hardness of the wood.

header   Also called a *lintel*. A horizontal member over a door, window, or other opening; supports the members above it. Usually made of wood, stone, or metal. Also, in the framing of floor or ceiling openings, beam used to support the ends of joists.

hearth   The floor and front extension of a fireplace or the fireproof floor beneath a wood-burning stove. Usually made of brick, stone, tile, or concrete.

heartwood   In a log, the wood between the pith and the sapwood. Usually full of resins, gums, and other materials that make it dark in color and resistant to decay.

heel   The end, or foot, of a rafter that rests on the top plate of a wall.

heel cut   The vertical cut at the end of a rafter; helps form the *bird's-mouth.*

hip   The convex angle formed by the meeting of two roof slopes. Typically set at a 90-degree angle to each other.

hip rafter   A rafter that runs from the corner of a wall to the ridge board and forms the hip. Set at a 45-degree angle to the walls.

hip roof   A roof or portion of roof that slopes up toward the center from all sides.

hot wire   In an electrical circuit, any wire that carries current from the power source to an electrical device. The hot wire is usually identified with black, blue, or red insulation, but it can be any color except white or green.

I beam   A steel team that resembles the letter I when seen in cross-section. Often used for long spans under a house or to support ceiling joists where no bearing wall or partition is wanted.

inlay   Any decorative piece set into the surface of another piece. In hardwood strip floors, a border of contrasting wood around the edges.

insulation   Any material that resists the conduction of heat, sound, or electricity.

insulation board   A structural building board made of coarse wood or cane fiber in ½- and $^{25}/_{32}$-inch thicknesses. It can be obtained in various size sheets in various densities, and with several treatments.

insured loan   A loan insured by FHA, VA, or FmHA or a private mortgage insurance company.

interior finish   Any material (wall coverings, trim, etc.) used to cover the framing members of the interior of a structure.

isolation joint   Any joint where two incompatible materials are mechanically separated to prevent a chemical or galvanic reaction between them. Also provides opportunity for shifting and settling of the earth without cracks occurring in the materials.

jack rafter   Any rafter that spans the space between a top plate and hip rafter or a valley rafter and ridge board.

jack stud   Any short stud that doesn't go all the way from the soleplate to the top plate. Also called a *cripple stud.*

jamb   The frame surrounding a door or window; consists of two vertical pieces called side jambs and a top horizontal piece called a head jamb.

jig   A device that serves as a guide or template for cutting or shaping several similar pieces.

joint   The junction where two or more pieces of material meet and are held together. Examples are a dado joint, dovetail joint, lap joint, and mortise-and-tenon joint.

joint compound   A powder mixed with water or a ready-mixed compound for application to the seams between sheets of wallboard.

jointing   The smoothing or straightening of the edges of boards so that they fit together precisely. A machine that performs this job is called a jointer.

joist   One of a series of parallel beams, usually two inches in thickness, used to support floor and ceiling loads and supported in turn by larger beams, girders, or bearing walls.

junction box   A metal or plastic container for electrical connections. Sometimes called a *box* or electrical box.

kerf   The cut made by a saw.

key   A small strip of wood inserted into one or both parts of a joint; aligns the pieces and holds them together. Also called a *spline*.

keyway   The slot or groove that holds the key.

kiln-dried lumber   Lumber that has been kiln-dried, often to a moisture content of 6 to 12 percent. Common varieties of softwood lumber, such as framing lumber, are dried to a higher moisture content.

kneewall   A short wall extending from the floor to the sloping ceiling of an attic or top-story room.

knockout   A die-cut impression in the wall of a junction box; can be removed (knocked out) to provide access for wires or cable.

laminate   To form a panel or sheet by bonding two or more layers of material. Also, a product formed by such a process—plastic laminate used for countertops, for example.

landing   The platform between flights of stairs or at the end of a stairway.

late charges   An additional fee charged if mortgage payments are not made on time.

lath or **lathe**   A building material of metal, gypsum, wood, or other material; used as a base for plaster or stucco.

lattice   Any framework of crossed slats of wood, metal, or plastic.

layout   Any drawing showing the arrangement of structural members or features. Also the act of transferring the arrangement to the site.

ledger   A board or strip of wood, fastened to the side of a wall or other framing member, on which other framing members (usually joists) rest or to which they are attached. Also called ledger board or ledger strip.

level   The position of a vertical line from any place on the surface of the earth to the center of the earth. Also, the horizontal position parallel to the surface of a body of still water. Also a device used to determine when surfaces are level or plumb.

lien   The right to hold a piece of property of a debtor as security for payment. In construction, subcontractors may file a mechanic's lien on a property if they are not paid for the work they performed on that property.

linear measure   Any measurement along a line.

lintel   A horizontal member over a door, window, or other opening; supports the members above it. Usually made of wood, stone, or metal. Also called a *header*.

live load   All loads on a building not created by the structure of the building itself; the furniture, people, and other things that occupy the building.

loan-to-value (LTV) ratio   The relationship between the amount of a mortgage loan and the value of the property.

lookout   A framing member that projects beyond the walls of a structure to support a roof overhang.

louver   One of a series of parallel slats arranged to permit ventilation and to limit or exclude light, vision, or weather. Louvers can be stationary or movable.

lumber   Wood product manufactured by sawing, resawing, and passing lengthwise through a planing machine, and then crosscutting to length.

main drain   In plumbing, the pipe that collects the discharge from branch waste lines and carries it to the outer foundation wall, where it connects to the sewer line.

main vent   In plumbing, the largest vent pipe to which branch vents may connect. Also called the vent stack.

mansard   A roof that slopes very steeply around the edge of a structure, providing room for a complete story. The central area of the roof inside the mansard roof is usually flat or slightly sloped. This kind of roof was named for the architect who designed it in order to avoid taxes when they were levied on all stories below the eaves.

masonry   Stone, brick, concrete, hollow tile, concrete block, gypsum, block, or other similar building units or materials bonded together with mortar to form a foundation, wall, pier, buttress, or similar mass.

mastic   A viscous material used as an adhesive for setting tile or resilient flooring.

matched lumber   Lumber that is dressed and shaped on one edge in a grooved pattern and on the other in a tongued pattern.

mortar   A mixture of sand and portland cement; used for bonding bricks, blocks, tiles, or stones.

mortgage   An agreement between a lender and a buyer using real property as security for the loan.

mortise and tenon   A joint made by cutting a mortise, or hole, in one piece and a tenon, or projection, to fit into the mortise on the other piece.

mudsill   The lowest member in the framing of a structure; usually 2-by (as in 2×4, 2×6, etc.) lumber bolted to the foundation wall on which the floor joists rest. Also called a *sill plate*.

mullion   The vertical divider between the windows in a window unit that is made up of two or more windows.

muntin   The dividers, either vertical or horizontal, that separate the small lights in a multipaned sash.

NM cable   Nonmetallic sheathed electric cable used for indoor wiring. Also known by the brand name Romex.

natural finish   A transparent finish—usually sealer, oil, or varnish—that protects the wood but allows the natural color and grain to show through.

neutral wire   In a circuit, any wire that is kept at zero voltage. The neutral wire completes the circuit from source to fixture or appliance to ground. The covering of neutral wires is always white.

newel   The main post at the foot of a stairway. Also the central support of a winding or spiral flight of stairs.

nipple   In plumbing, any short length of pipe externally threaded on both ends.

nominal size   The size designation of a piece of lumber before it is planed or surfaced. If the actual size of a piece of surfaced lumber is $1\frac{1}{2} \times 3\frac{1}{2}$ inches, it is referred to by its nominal size: 2×4.

nonbearing wall   A wall supporting no load other than its own weight.

nosing   The part of a stair tread that projects over the riser. Also the rounded edge on any board.

note   A loan agreement.

o.c.   Abbreviation for on center; the measurement of spacing for studs, rafters, joists, and posts from the center of one member to the center of the next.

on center   Referring to the spacing of joists, studs, rafters, or other structural members as measured from the center of one to the center of the next. Usually written o.c. or OC.

oriented strand board (O.S.B)   A panel material of wood flakes compressed and bonded together with phenolic resin. Used for many of the same applications as plywood. Also known as structural flakeboard.

outlet In a wall, ceiling, or floor, a device into which the plugs on appliance and extension cords are placed to connect them to electric power. Properly called a *receptacle*.

outrigger An extension of a rafter or a small member attached to a rafter that forms a cornice or overhand on a roof.

panel A large, thin board or sheet of construction material. Also a thin piece of wood or plywood in a frame of thicker pieces, as in a panel door or wainscoting.

panel siding Large sheets of plywood or hardboard that may serve as both sheathing and siding on the exterior of a structure.

parquet A type of wood flooring in which small strips of wood are laid in squares of alternating grain direction. Parquet floors are now available in ready-to-lay blocks to be put down with mastic. Also any floor with an inlaid design of various woods.

particleboard A form of composite board or panel made of wood chips bonded with adhesive.

partition A wall that subdivides any room or space within a building.

penny As applied to nails, it originally indicated the price per hundred. The term now serves as a measure of nail length and is abbreviated by the letter *d*.

phillips head A kind of screw and screwdriver on which the diving mechanism is an X rather than a slot.

pier A column of masonry, usually rectangular, used to support other structural members. Often used as a support under decks.

pigtail A short length of electrical wire or group of wires.

piling Long post of metal or moisture-resistant wood driven into the ground wherever it is difficult to secure a firm foundation in the usual way, usually in soft or swampy areas.

pitch The incline of a roof.

PITI An acronym for principal, interest, taxes, and insurance.

plan The representation of any horizontal section of a structure, part of a structure, or the site of a structure; shows the arrangement of the parts in relation and scale to the whole.

plank A broad piece of lumber, usually more than one inch thick, especially one used to stand on as part of a scaffold or between ladders or sawhorses.

plaster A mixture of lime, sand, and water plus cement for exterior cement plaster, and plaster of paris for interior smooth plaster used to cover the surfaces of a structure.

plasterboard *See* wallboard.

plate A horizontal framing member, usually at the bottom or top of a wall or other part of a structure, on which other members rest. The mudsill, soleplate, and top plate are examples.

plate cut In a rafter, the horizontal cut that forms the bird's-mouth. Also called a seat cut.

platform framing A system of framing a structure in which the floor joists of the first story rest on the mudsill of the foundation, and those of each additional story rest on the top plates of the story below it. All bearing walls and partitions rest on the subfloor of their own story.

plug In plywood, a piece of wood put in to replace a knot. On the cord of an appliance, the device that inserts into a receptacle.

plumb Exactly perpendicular; vertical.

ply A term used to refer to a layer in a multilayered material, such as one layer of a sheet of plywood.

plywood   A wood product made up of layers of wood veneer bonded together with adhesive. It is usually made up of an odd number of plys set at a right angle to each other.

points   Equal to 1 percent of the loan amount; normally charges for various services or fees collected as an up-front cost in addition to the down payment.

porch   A floor extending beyond the exterior walls of a structure. It may be enclosed, covered, or open.

post   A vertical support member, usually made up of only one piece of lumber or a metal pipe or I beam.

post-and-beam   A method of construction that, rather than stud walls, uses beams spanning between posts as the main support structure.

prepayment privileges   A loan option that enables the borrower to repay all or part of the loan in advance without incurring penalty.

prequalification   A process in which a potential home buyer qualifies for a mortgage before making an offer on a house.

preservative   Any substance that, for a reasonable length of time, will prevent the action of wood-destroying fungi, borers of various kinds, and similar destructive agents when the wood has been properly coated or impregnated with it.

principal and interest   The monthly cost of a mortgage; principal is the amount borrowed—the difference between the cost of the home and the down payment—while interest is the charge made by the lender for lending the money.

program   A written list of requirements to be included in the design of a building.

punch list   A written list of unfinished or incorrectly finished items that must be finished or corrected before the owner will accept the project from the contractor.

purlin   A horizontal framing member that supports common rafters in a roof. Usually the board between the slopes of a gambrel roof.

putty   A soft, pliable material used for sealing the edges of glass in a sash or to fill small holes or cracks in wood.

PVC (polyvinyl chloride)   A rigid white plastic pipe used in plumbing for supply and DWV systems.

quarter-round   A convex molding shaped like a quarter circle when viewed in cross section, typically used as wall trim at the floor or ceiling.

rabbet   A rectangular channel cut in the corner of a piece of lumber.

radiant heating   Electrically heated panels or hot-water pipes in the floor or ceiling that radiate heat to warm the room's surfaces.

rafter   One of a series of parallel framing members, usually made from 2-by lumber, that support a roof.

rail   In a balustrade, stairway, or fence, the horizontal or slanted member extending from the top of one post or support to another. Also the bottom member of a balustrade parallel to the top rail. Also the horizontal members of a door or window frame.

rake   The inclined edge of a gable roof. The trim piece on the rake is called a rake molding.

receptacle   In a wall, ceiling, or floor, an electric device into which the plugs on appliance and extension cords are placed to connect them to electric power. Also called an *outlet*.

register   In a wall, floor, or ceiling, the device through which air from the furnace or air conditioner enters a room. Also any device for controlling the flow of heated or cooled air through an opening.

reinforcing   Steel bars or wire mesh placed in concrete to increase its strength.

repayment period   The number of years it will take to repay the loan.

resawing   Sawing lumber again, after the first sawing, to make it an unusual size or to shape it, as with bevel siding.

retainage   The amount (usually 10 percent) held back by an owner out of each payment to the general contractor, to be held as security that the work will be finished and to be paid when the work is complete.

ribbon   A narrow board let into studs to support joists or beams, usually in balloon framing.

ridge   The horizontal line where two roof slopes meet. Usually the highest place on the roof.

ridge board   The board placed at the ridge of the roof to which the upper ends of the rafters are attached.

ridge cut   The vertical cut at the upper end of a rafter.

rim joist   A joist nailed across the ends of the floor or ceiling joists. Also known as a *band joist*.

ripping   Sawing wood in the direction of the grain.

rise   In a stairway, the vertical measurement from one tread to the next. In a roof, the vertical measurement from the top of the doubled top plate to the top of the ridge board.

riser   Each of the vertical boards between the treads of a stairway.

roll roofing   Material made of asphalt-saturated fiber and coated with mineral granules. Typically sold in 36-inch rolls.

Romex   A brand name for nonmetallic sheathed electric cable used for indoor wiring. Also called Type NM cable.

roof   The uppermost part of a structure that covers and protects it from weather. Also the covering on this part of a structure.

roof sheathing   Fastened to the rafters, sheet material, or boards to which the roofing is applied.

roofing   The material put on the roof to make it impervious to the weather.

rough lumber   Lumber as it comes from the saw, before it is surfaced by a planer.

rough-in   To install the basic, hidden parts of a plumbing, electrical, or other system while the structure is in the framing stage. Contrasts with installation of finish electrical work or plumbing, which consists of the visible parts of the system.

run   In stairways, the front-to-back width of a single stair or the horizontal measurement from the bottom riser to the back of the top tread.

saddle   A U-shaped metal or plastic flashing installed on a roof above or below a chimney, skylight, or other roof protrusion to divert water. Also a single-or-double-sloped structure, usually made of wood, placed on a roof to divert water from between two surfaces that meet at an angle.

sand float finish   Lime mixed with sand, resulting in a textured finish.

sanding   Rubbing sandpaper or similar abrasive material over a surface to smooth it or prepare it for finishing.

sapwood   The wood near the outside of a log; contains the living cells of a tree. Usually lighter in color than heartwood, and more susceptible to decay and termite infestation.

sash   The frame that holds the glass panes in a window.

saw kerf   General term for the cut made by a saw.

scaffold   A temporary structure to support workers when they are working on parts of a building higher than they can reach from the ground or floor.

scale   The proportion between two sets of dimensions. On building plans, the house is drawn smaller than the actual house, but in scale so that the proportions are the same. For example, when the scale is expressed as ¼" = 1'0", ¼ inch on the drawing equals 1 foot on the actual house.

screed   A strip of wood used to even the surface and control the thickness of gravel beneath a foundation slab or the various layers of plaster or stucco on walls or ceilings or to strike off wet cement.

scribing   Making a piece of wood or paneling so that it can be cut to fit precisely against an irregular surface. Also the cutting of the scribed line.

sealer   A finishing material used to seal the surface of wood or other material before the final coats of paint or stain are applied.

seasoning   Removing moisture from green lumber either by air drying or kiln drying.

second mortgage   A loan in which the lender has secondary rights to the property.

section   A drawing or part of a building as it would appear if cut through by a vertical plane.

sepias   Prints of original drawings that are on translucent paper and may be used to make more prints.

service entrance   The place where electrical utilities enter a building.

service panel   The box or panel where the electricity is distributed to the house circuits. It contains the circuit breakers and, usually, the main disconnect switch.

shake   A thick wooden shingle, usually edge-grained.

shared appreciation mortgage   A loan that enables the lender and the buyer to share any increase in value or a property.

shared equity loan   A mortgage that makes it possible for a relative or investor to help with the down payment for a share in the equity of the house.

sheathing   Sheet material or boards fastened to the rafters or exterior stud walls; that to which the roofing or siding is applied.

shed roof   A roof that slopes in only one direction.

sheet metal work   Any components of a structure in which sheet metal is used, such as ducts and flashing.

sheetrock   A commercial name for wallboard.

shim   A thin wedge of wood, often part of a shingle, used to bring parts of a structure into alignment.

shingles   A roofing material of asphalt, fiberglass, wood, or other material cut to stock lengths, widths, and thicknesses.

shiplap siding   Exterior siding, usually ¾ inch thick, shaped to patterns.

shoe molding   A strip of wood used to trim the bottom edge of a baseboard.

shutoff valve   In plumbing, a fitting to shut off the water supply to a single fixture or branch of pipe.

shutters   Flush-board, louvered, or paneled frames in the form of small doors that are mounted alongside a window. Some are made to close over the window to provide protection or privacy; others are nailed or screwed to the wall for decorative purposes only.

siding   The finish covering on the exterior walls of a building.

sill plate   The lowest member in the framing of a structure; usually a 2-by board bolted to the foundation wall on which the floor joists rest. Also called a *mudsill*.

site plan   Drawing of all the existing conditions on the lot, usually including slope and other topography, existing utilities, and setbacks. These drawings may be provided by the municipality.

slab A concrete foundation or floor poured directly on the ground.

sleepers Boards embedded in or attached to a concrete floor; serve to support and provide a nailing surface for a subfloor or finish flooring.

slope The incline or pitch of a roof.

soffit The underside of a stairway, cornice, archway, or similar member of a structure. Usually a small area relative to a ceiling.

soft costs All the costs associated with the beginning of a construction project that purchase intangible items that cannot be resold, such as legal fees, architect and engineering fees, loan points, and surveys.

softwood The wood of conifers such as fir, pine, and redwood. Softwood is usually softer than hardwood, but the term has no actual reference to the hardness of the wood.

soil cover A light covering of plastic film, roll roofing, or similar material used over the soil in crawl spaces of buildings to minimize moisture permeation of the area. Also called ground cover.

soil stack In the DWV system, the main vertical pipe. Usually extends from the basement to a point above the roof.

solderless connector A product that establishes connection between two or more electrical conductors without solder. Also called a wire nut.

soleplate In a stud wall, the bottom member, which is nailed to the subfloor. Also called a bottom plate.

solid bridging A solid member placed between adjacent floor joists near the center of the span to prevent joists from twisting.

span The distance between structural supports, such as walls, columns, piers, beams, girders, and trusses.

specifications Written lists, instructions, and general information that relate to the construction and make up a part of the total legal contract.

splash block A small masonry block laid with the top close to the ground surface to receive roof drainage from downspouts and to carry it away from the foundation.

spline A small strip of wood inserted into one or both parts of a joint; aligns the pieces and holds them together. Also called a *key*.

square A term used to describe an angle of exactly 90 degrees. Also a device to measure such an angle. Also a unit of measure equaling 100 square feet, usually used in describing amounts of roofing or siding material.

stile A vertical framing member in a panel door, wainscoting, or a paneled wall.

stool The horizontal shelf on the interior of the bottom of a window.

story The part of a building that is between floors or the top floor and the roof.

stringer In a stairway, the supporting member to which the treads and risers are fastened. Also called a carriage.

strip flooring Wood flooring consisting of narrow, matched strips.

stucco A plaster of sand, portland cement, and lime used to cover the exterior of buildings.

stud One of a series of wood or metal vertical framing members that are the main units of walls and partitions.

stud wall The main framing units for walls and partitions in a building, composed of studs; top plates; bottom plates; and the framing of windows, doors, and corner posts.

subfloor   Plywood or oriented strand boards attached to the joists. The finish floor is laid over the subfloor.

substantial completion   The time at which the contractor feels that he or she has essentially finished the project, but before the final inspection.

suspended ceiling   A system for installing ceiling tile by hanging a metal framework from the ceiling joists.

switch   In electrical systems, a device for turning the flow of electricity on and off in a circuit or diverting the current from one circuit to another.

tail cut   At the lower end of a rafter, the vertical cut to which the fascia board is attached.

tail joist   A relatively short joist or beam, usually used to frame an opening in a floor or ceiling; supported by a wall and a header at the other.

taper   A uniform decrease in size from one end to the other—as of a table leg, for instance.

template   A pattern from which parts of a structure can be made. Templates may be of paper, cardboard, plywood, and so forth.

tenon   A projection of one piece of wood that fits precisely into a hole, the mortise, in another piece to form a mortise-and-tenon joint.

termite shield   Galvanized steel or aluminum sheets placed between the foundation, pipes, or fences and the wood structure of a building; prevents the entry of termites.

threshold   A shaped piece of wood or metal, usually beveled on both edges, that is placed on the finish floor between the side jamb; forms the bottom of an exterior doorway.

tie beam   *See* collar beam.

timber   Pieces of lumber with a cross section greater than four by six inches. Usually used as beams, girders, posts, and columns.

title   A document that gives evidence of property ownership.

toenailing   Driving a nail at a slant to the initial surface in order to permit it to penetrate into a second member.

tongue   A projecting edge on a board; fits into a groove on another board.

tongue and groove   A way of milling lumber so that it fits together tightly and forms an extremely strong floor or deck. Also boards milled for tongue-and-groove flooring or decking that have one or more tongues on one edge and a matching groove or grooves on the other.

top plate   In a stud wall, the top horizontal member to which the cap plate is nailed when the stud walls are connected and aligned.

transom   A window above a doorway.

trap   In plumbing, a U-shaped drain fitting that remains full of water to prevent the entry of air and sewer gas into the building.

tread   In a stairway, the horizontal surface on which a person steps.

trim   Any finish materials in a structure that are placed to provide decoration or to cover the joints between surfaces or contrasting materials. Door and window casings, baseboards, picture moldings, and cornices are examples of trim.

trimmer   A joist or beam to which a header is nailed, usually when framing an opening in a floor or ceiling. Also a stud next to an opening in a wall that supports a header.

truss   A frame or jointed structure designed to act as a beam of long span, while each member is usually subjected to longitudinal stress only, either tension or compression.

truss plate   A flat piece of metal, wood, or plywood, used to support the connections of framing lumber. Most often seen on joints of wood trusses and joist splices. Also called a *gusset*.

Truth-In-Lending Act   A federal law that states that lenders must reveal all the terms of the mortgage.

underlayment   The material placed under the finish coverings of roofs or floors to provide waterproofing as well as a smooth, even surface on which to apply finish material.

underwriting   The process of evaluating a mortgage application to determine the risk to the lender; involves an analysis of the value of the property and the borrower's ability to make payments.

unit of rise   A division of the total vertical height of a roof slope or a stairway; used in calculating the length of rafters and the relationship of risers to treads.

unit of run   A portion of the total horizontal length of a roof slope or a stairway, used in calculating the length of rafters and the relationship of the risers to treads.

VA   Veterans Administration; guarantees loans for qualified veterans and spouses.

valley   The concave angle formed by the meeting of two sloping surfaces of a roof, that come off adjacent walls to form an inside corner.

valley flashing   A method of waterproofing the valley of a roof with metal or roofing-felt flashing.

valley rafter   A rafter that runs from a wall plate at the corner of the house, along the roof valley, and to the ridge.

vapor barrier   Any material used to prevent the penetration of water vapor into walls or other enclosed parts of a building. Polyethylene sheets, aluminum foil, and building paper are the materials used most.

veneer   A thin layer of wood, usually one that has beauty or value, that is applied for economy or appearance on top of an inferior surface.

vent   Any opening, usually covered with screen or louvers, made to allow the circulation of air, usually into an attic or crawl space. In plumbing, a pipe in the DWV system for the purpose of bringing air into the system.

vent stack   In plumbing, the largest vent pipe to which branch vents may connect. Also called the *main vent*.

wainscoting   Panel work covering only the lower portion of a wall.

wall plate   A decorative covering for a switch, receptacle, or other device.

wallboard   Panels consisting of a layer of gypsum plaster covered on both sides with paper, used for finishing interior walls and ceilings. Also called gypsum wallboard, drywall, and Sheetrock.

water-repellent preservative   A liquid designed to penetrate into wood and impart water repellency and a moderate preservative protection.

weather stripping   Narrow strips of metal, fiber, plastic foam, or other materials placed around doors and windows; prevents the entry of air, moisture, or dust.

wire nut   A device that uses mechanical pressure rather than solder to establish a connection between two or more electrical conductors. Also called a *solderless connector*.

Appendix B

Home Construction Resources

This book intends to answer your primary questions about how to build your own home, offering specific options and processes. For additional information on designing, planning, financing, building, and enjoying your home, refer to the nearly 200 resources in this appendix.

The Internet is the best source of current information. It is also widely available in homes, businesses, libraries, and even coffee shops. Therefore, resources in this appendix refer you to websites that will include up-to-date general and specific information on a wide variety of building topics. Remember to add the prefix "www." to all URLs in this appendix.

Home Building Trade Association Resources

Air Conditioning Contractors of America: acca.org

American Concrete Institute: aci-int.org

American Concrete Pavement Association: pavement.com

American Consulting Engineers Council: acec.org

American Forest & Paper Association: afandpa.org

American Hardware Manufacturers Association: ahma.org

American Institute of Architects: aiaonline.com

American Institute of Building Design: aibd.org

American Institute of Steel Construction: aisc.org

American Institute of Timber Construction: aitc-glulam.org

American Iron & Steel Institute: steel.org

American Lighting Association: americanlightingassoc.com

American National Standards Institute: ansi.org

American Society for Testing and Materials: astm.org

American Society of Civil Engineers: asce.org

American Society of Heating, Refrigerating & Air Conditioning Engineers: ashrae.org

American Society of Interior Designers: asid.org

American Solar Energy Society: sni.net/solar

American Water Works Association: awwa.org

American Wind Energy Association: igc.apc.org/awea

American Wood Preservers Institute: awpi.org

Americans with Disabilities Act: usdoj.gov/crt/ada/adahom1.htm

Architectural Woodwork Institute: awinet.org

Asphalt Institute: asphaltinstitute.org

Asphalt Roofing Manufacturers Association: asphaltroofing.org

Associated Builders and Contractors: abc.org

Associated General Contractors of America: agc.org

Associated Soil and Foundation Engineers: asfe.org

Association of Home Appliance Manufacturers: aham.org

Brick Institute of America: bia.org

Building Officials and Code Administrators International: bocai.org

California Redwood Association: calredwood.org

Canadian and American Log Builders' Association, International: woodworking.com/loghomes/logassoc/index.html

Canadian Homebuilders' Association: chba.ca

Canadian Institute of Plumbing and Heating: ciph.com

Canadian Standards Association: csa.ca

Canadian Window & Door Manufacturers Association: windoorweb.com

Cast Iron Soil Pipe Institute: cispi.org

Cedar Shake & Shingle Bureau: cedarbureau.org

Cellulose Insulation Manufacturers Association: cellulose.org

Composite Panel Association and Composite Wood Council: pbmdf.com

Concrete Reinforcing Steel Institute: crsi.org

Construction Specifications Institute: csinet.org

Contractor License Reference Site: contractors-license.org

The Council of American Building Officials: cabo.org

Energy Efficient Building Association: eeba.org

Engineered Wood Association: apawood.org

Gypsum Association: gypsum.org

Hardwood Council: hardwoodcouncil.com

Hardwood Plywood & Veneer Association: erols.com/hpva

Home Builders Institute: hbi.org

Institute of Electrical and Electronics Engineers: ieee.org

Insulating Concrete Form Association: forms.org

International Conference of Building Officials: icbo.org

International Solar Energy Society: ises.org

International Standard Organization: iso.ch

Kitchen Cabinet Manufacturers Association: kcma.org

Light Gauge Steel Engineers Association: lgsea.com

Log Homes Council: loghomes.org

Manufactured Housing Institute: mfghome.org

National Air Duct Cleaners Association: nadca.com

National Asphalt Pavement Association: hotmix.org

National Association of Home Builders: nahb.com

National Association of Women in Construction: nawic.org

National Concrete Masonry Association: ncma.org

National Electrical Contractors Association: necanet.org

National Electrical Manufacturers Association: nema.org

National Fire Protection Association: nfpa.org

National Hardwood Lumber Association: natlhardwood.org

National Institute of Building Sciences: nibs.org

National Institute of Standards and Technology: nist.gov

National Oak Flooring Manufacturers Association: nofma.org/index.htm

National Pest Control Association: pestworld.org

National Roofing Contractors Association: roofonline.org

National Rural Water Association: nrwa.org

National Sash & Door Jobbers Association: nsdja.com

National Stone Association: aggregates.org

National Tile Contractors Association: tile-assn.com

National Tile Roof Manufacturers Association: ntrma.com

National Wood Flooring Association: woodfloors.org

National Wood Window and Door Association: nwwda.org

North American Insulation Manufacturers Association: naima.org

North American Steel Framing Alliance: steelframingalliance.com

Occupational Safety and Health Administration: osha.gov

Passive Solar Industries Council: psic.org

Plumbing, Heating, Cooling Contractors National Association: naphcc.org

Portland Cement Association: concretehomes.com

Precast/Prestressed Concrete Institute: pci.org

Quartzite Rock Association: quartzite.com

Sheet Metal and Air Conditioning Contractors' National Association: smacna.org

Single Ply Roofing Institute: spri.org/index.html

Southern Building Code Congress International: sbcci.org

Southern Forest Products Association: sfpa.org

Southern Pine Council: southernpine.com

Steel Door Institute: wherryassoc.com/steeldoor.org/default.html

Structural Board Association: sba-osb.com

Timber Framers Guild of North America: tfguild.org

Underwriters Laboratories Inc.: ul.com

Vinyl Siding Institute: vinylsiding.org

Western Red Cedar Lumber Association: cofi.org/WRCLA

Western Wood Products Association: wwpa.org

Wood Floor Covering Association: wfca.org

Government and Financial Resources

Department of Housing and Urban Development: hud.gov

Environmental Protection Agency: epa.gov

Federal Home Loan Mortgage Corporation: freddiemac.com

Federal Housing Administration: hud.gov/fha

Federal Housing Finance Board: fhfb.gov

Federal National Mortgage Association: fanniemae.com

Small Business Administration: sba.gov

Veterans Administration Home Loan Guarantee Program: homeloans.va.gov

Owner-Builder Schools

Arcosanti: arcosanti.org

Building Education Center: bldgeductr.org

Earthwood Building School: interlog.com/~ewood

Fox Maple School of Traditional Building: foxmaple.com

Heartwood School for the Homebuilding Crafts: heartwoodschool.com

Not So Big House: notsobighouse.com

Owner-Builder Dome School: naturalspacesdomes.com/schoolpi.htm

Phinney Well Home Program: phinneycenter.org

Shelter Institute: shelterinstitute.com

Southface Energy Institute: southface.org

Yestermorrow Design/Build School: yestermorrow.org

Books and Videos on Building and Maintaining Houses

These books are available through local bookstores or www.MulliganBooks.com.

Allen, Edward, et. al. *Fundamentals of Building Construction*. John Wiley, 1998.

Ballard, Scott T. *The Complete Guide to Designing Your Own Home*. Betterway, 1995.

Ching, Frank. *Building Construction Illustrated*. John Wiley, 2000.

DiDonno, Lupe, and Phyllis Sperling. *How to Design & Build Your Own Home*. Alfred A. Knopf, 1987.

Folds, John, and Ray Hoopes. *Building the Custom Home*. Taylor, 1990.

Jones, Jack P. *Handbook of Construction Contracting* (two volumes). Craftsman, 1987 and 1989.

Kardon, Redwood, et. al. *Code Check: A Field Guide to Building a Safe House*. Taunton Press, 2000.

Managing Home Construction (two-tape video) with Dean Johnson and Robin Hartl. Hometime.com, 1994.

McGuerty, Dave, and Kent Lester. *The Complete Guide to Contracting Your Home*. Betterway, 1997.

Miller, Mark R., Rex Miller, and Glen E. Baker. *Carpentry & Construction, Third Edition*. McGraw-Hill, 1999.

Nash, George. *Do-It-Yourself Housebuilding*. Sterling, 1995.

National Construction Estimator. Craftsman, 2001.

National Electrical Code. National Fire Protection Association, 2001.

National Standard Plumbing Code. National Association of Plumbing-Heating-Cooling Contractors, 2000.

Ramsey, Dan. *Builder's Guide to Barriers*. McGraw-Hill, 1996.

———. *Builder's Guide to Foundations and Floor Framing*. McGraw-Hill, 1995.

———. *Building a Log Home from Scratch or Kit*. TAB, 1983 and 1987.

———. *Hardwood Floors: Installing, Maintaining, and Repairing*. TAB/McGraw-Hill, 1985 and 1991.

———. *The Complete Idiot's Guide to Solar Power for Your Home*. Alpha Books, 2003.

———. *Tile Floors: Installing, Maintaining, and Repairing*. TAB/McGraw-Hill, 1985 and 1991.

———. *What the "Experts" May Not Tell You About Buying a House or Apartment*. Warner Books, 2004.

Residential Structure & Framing by *The Journal of Light Construction*. Craftsman, 1999.

Roskind, Robert and the Owner Builder Center. *Building Your Own House*. Ten Speed Press, 1984.

Seddon, Leigh. *Practical Pole Building Construction*. Williamson Publishing, 1985.

Tenenbaum, David J. *The Complete Idiot's Guide to Home Repair and Maintenance Illustrated*. Alpha Books, 2004.

———. *The Complete Idiot's Guide to Simple Home Improvements Illustrated*. Alpha Books, 2004.

Thallion, Rob. *Graphic Guide to Frame Construction*. Taunton Press, 2000.

Home and Construction Magazines

Affordable Housing Finance: housingfinance.com

American Bungalow: ambungalow.com

Automated Builder: automatedbuilder.com

Backwoods Home: backwoodshome.com

Builder: builderonline.com

Builder & Developer: bdmag.com

Builder/Architect: builderarchitect.com

Building Design & Construction: bdcmag.com

Canadian House & Home: canadianhouseandhome.com

Canadian Living: canadianliving.com

Coastal Living: coastallivingmag.com

Concrete Homes: concretehomesmagazine.com

Construction: construction.com

Construction Monthly: constmonthly.com

Contractor: contractormag.com

Country Living: countryliving.com

Environmental Building News: ebuild.com

Environmental Design & Construction: edcmag.com

Fine Homebuilding: finehomebuilding.com

Good Housekeeping: goodhousekeeping.com

Home and Design: homeanddesign.com

Home Systems: gohomesystems.com

House Beautiful: housebeautiful.com

Journal of Light Construction: jlconline.com

Log Home Design Ideas: lhdi.com

Log Home Living: homebuyerpubs.com

Manufactured Housing Today: mhtoday.com

Metal Construction News: moderntrade.com

Natural Home: naturalhomemagazine.com

Permanent Buildings & Foundations: pbf.org

Professional Builder: probuilder.com

Residential Architect: residentialarchitect.com

Shelterforce: nhi.org/online/sf.htm

Southern Living: southern-living.com

Sunset: sunsetmagazine.com

This Old House: thisoldhouse.com

Timber Frame Homes: homebuyerpubs.com

Traditional Building: traditionalbuilding.com

Wood Design & Building: wood.ca

Other Useful Resources

Bob Vila Online: www.bobvila.com

Design Works Home Planner: homeplanner.com

Fix-It Club: www.fixitclub.com

Free software and shareware: freeware.com and shareware.com

Garlinghouse Home Plans: garlinghouse.com

Google Search Engine: google.com

Grandy Post and Beam Homes: postbeam.com

Home Plan Finder: homeplanfinder.com

Mulligan Books: mulliganbooks.com

Index